The International Library of Bioethics

The *International Library of Bioethics* comprises volumes with an international and interdisciplinary focus on foundational and applied issues in bioethics. By means of this Series we aim to meet the challenge of our time: how to aim biotechnology to good human and other living things' ends, how to deal with changed values in the areas of religion, society, and culture, and how to formulate a new way of thinking, a new bioethic.

This Series was conceived against the background of increasing globalization and interdependency of the world's cultures and governments, with mutual influencing occurring throughout the world in all fields. We welcome book proposals representing the broad interest of this series' interdisciplinary and international focus. We especially encourage proposals that address aspects of changes in biological and medical research and clinical health care, health policy, medical and biotechnology, and other applied ethical areas involving living things, with an emphasis on those interventions and alterations that force us to re-examine foundational issues.

More information about this series at http://www.springer.com/series/16538

Pekka Louhiala

Placebo Effects: The Meaning of Care in Medicine

 Springer

Pekka Louhiala
University of Helsinki
Helsinki, Finland

University of Tampere
Tampere, Finland

ISSN 2662-9186   ISSN 2662-9194   (electronic)
The International Library of Bioethics
ISBN 978-3-030-27327-9   ISBN 978-3-030-27329-3   (eBook)
https://doi.org/10.1007/978-3-030-27329-3

This Springer imprint is published by the registered company Springer Nature Switzerland AG
The registered company address is: Gewerbestrasse 11, 6330 Cham, Switzerland

Preface

I do not remember exactly when and how I became interested in placebos and placebo effects. The first paper on these topics was a short essay for the Finnish Medical Journal in 1995. When I look at it now, I feel slightly embarrassed. In that essay, I made the same common mistake that I criticise in this book: I did not make a clear distinction between the *use of placebos* and *placebo effects*. Maybe this observation is a sign of progress. And if I'm alive after 20 years, I will probably feel embarrassed for some parts of this book as well.

One discussion I remember very clearly took place in a seminar about 15 years ago. I have forgotten the main topic of the seminar, but somehow we ended up in a discussion about placebo effects. My old friend and colleague Raimo Puustinen commented that the essence of a placebo effect is the patient's *experience of being cared for*. This was not an exact definition of a placebo effect, but I realised immediately that he had said something very important.

During the following years, Raimo and I have had numerous discussions about placebos and placebo effects, often ending in agreement but sometimes also in disagreement. The idea of replacing the term placebo effect with *care effect* came originally from Raimo, and we developed it together in a few scientific publications. The title of this book is based on Raimo's suggestion as well.

I am also grateful for many other people for inspiration and challenge. Martyn Evans, Harri Hemilä, Eija Kalso, Ted Kaptchuk, Hasse Karlsson and Daniel Moerman have in different ways contributed to the process that led to this book. Numerous students and colleagues have commented on my lectures or seminars on these issues, and Signe and Ane Gyllenberg Foundation has supported my work. I'm grateful to them.

Last year my father found a cardboard box that had been untouched in his home for decades. Among other interesting things, the box contained a set of offprints of my uncle Heikki Kalliola's medical publications from the late 1950s and the early

1960s. One of the publications was titled "Placebo effect in peptic ulcer and other gastrointestinal disorders". My uncle had passed away the year before, and I never had the chance to discuss these things with him. The moment I saw his paper was, however, touching. The story continues.

Hämeenlinna, Finland   Pekka Louhiala
June 2019

Part of the Text Builds Upon the Following Papers, and Both the Co-authors and the Publishers Have Given Their Permission for the Use of the Papers

Louhiala P, Puustinen R. Rethinking the placebo effect. Medical Humanities 2008;34:107–109.

Louhiala P. The ethics of the placebo in clinical practice revisited. J Med Ethics 2009;35:407–409.

Louhiala P. What do we really know about the deliberate use of placebos in clinical practice? J Med Ethics 2012;38:403–405.

Louhiala P, Puustinen R. Placebo effect or care effect? Four examples from the literary world. Hektoen International 2013 volume 5, issue 4 http://hekint.org/2017/03/03/placebo-effect-or-care-effect-four-examples-from-the-literary-world/?highlight=louhiala. Accessed 1 Jun 2018.

Louhiala P, Hemilä H, Puustinen R. Impure placebo is a useless concept. Theoretical Medicine and Bioethics 2015;36:279–289.

Contents

Chapter 1
Introduction

The placebo effect is a magical reality underlying so much of our medical remedies, scientific or alternative. (Nayernouri 2017)

Placebo is about the attention, the eye gaze, the warmth, the compassion, the confidence in the doctor-patient-relationship. I would say that the placebo is about the symbols in medicine, like a diploma on the wall, the prescription pad, the stethoscope, …, about the routine rituals, procedures in medicine, waiting at the doctor's office, talking, disrobing, being examined, putting back your clothes, getting a diagnosis and then being prescribed pills, injections or procedures. Ultimately, I think the placebo effect is the imagination, hope and trust in the clinical encounter. (Kaptchuk 2012)

People respond to what we know, think, and feel…

People respond to what we are told, believe and know…

People respond to their various cultural backgrounds…

They respond to language, to caring, to culture, to community, to history. In a word, they respond to meaningful phenomena.

(Moerman 2013).

Many years ago I had a common cold with exceptional tiredness and occasional irregular pulse. This led me to consult a doctor, who turned out to be my former student. After a careful clinical examination and studying the electrocardiogram (ECG) the doctor explained that the results were normal and there were only a few benign extrasystoles seen in the ECG. Suddenly, the tiredness went away and I felt well again (Puustinen and Louhiala 2013).

A good friend of mine, also a doctor, was once working in an emergency ward when a middle-aged man with a severe pain in his left flank was rushed in. It turned out that the patient's pain was due to a kidney stone. My friend told the nurse to get injectable indomethacin. The nurse returned with a syringe filled with opaque liquid. The physician confirmed that the liquid was the substance required and the nurse nodded her head. The physician then injected the liquid into the patient's vein and assured him that the pain would disappear within a minute or two. Soon the patient's

© Springer Nature Switzerland AG 2020
P. Louhiala, *Placebo Effects: The Meaning of Care in Medicine*,
International Library of Ethics, Law, and the New Medicine 81,
https://doi.org/10.1007/978-3-030-27329-3_1

expression relaxed, he grasped the physician's arm and thanked him with tears in his eyes. The pain was gone. My friend went to the office to write notes. There was an empty box of injectable indomethacin on the table. When the physician dropped the box into the bin he noticed a small bottle in the box containing white powder. He realised that the inexperienced young nurse had not mixed the indomethacin powder into the saline he had injected into the patient (Puustinen and Louhiala 2013).

In a well-known paragraph of Book 11 of his Confessions, St. Augustine ponders the nature of time and asks: "What then is time? If no one asks me, I know: if I wish to explain it to one that asketh, I know not." (St. Augustine 1838). Science has proceeded since the 4th century but many concepts remain as obscure as they were then. "Placebo" and "placebo effect" entered into the scientific language much later but in many senses they remain enigmatic.

Physicians, scientists and lay people all know *something* about placebos. When asked, they would usually define placebos as *fake* or *ineffective* substances used in medical research and perhaps sometimes in the treatment of patients. They know also about placebo effects but defining these is far more difficult: if something is, by nature, ineffective, how can it have an effect? It is, however, commonly thought that placebo effects do exist and have some relevance in medical research and practice.

However, many issues related to placebos and placebo effects keep puzzling physicians, patients, and the scientific community. Since nothing in nature is, in fact, completely ineffective, how could placebos actually exist? And if they do exist, how could there any 'placebo effect' at all? If placebo effects are real, are they clinically relevant? Do placebos have a place in modern medicine? What does the existence of placebo effects tell us about the nature of healing?

One of the central aims of this book is to make sense of the concepts placebo and placebo effect. To introduce the reader to the field I shall, however, first leave the conceptual problems aside and present ten studies from over four decades. They all illustrate the field from different perspectives, from neuroscience to meta-analysis, from case studies to clinical research and from epidemiology to anthropology. The idea of ten papers was inspired by Daniel Moerman's talk in Helsinki in 2012 and his later article (Moerman 2013).

After presenting the studies I'll leave the world of science for a while and dwell into the world of literature. While the term placebo effect is not widely used in world literature, even fictional doctors have often been depicted as knowing that their patients may require no active drugs and that their mere presence, their advice and encouragement, will often lead to improvement of their symptoms or even recovery from their illness.

In the last part of this chapter I shall return to the concepts and introduce the reader to the conceptual confusion around them. After finishing the first chapter the reader may be slightly confused, but that is my purpose: fixing a problem begins by recognising that there is a problem.

1.1 Ten Papers—Ten Perspectives

The following collection of ten papers is a personal selection of scientific articles that have, in my opinion, changed the world or at least paved the way for new lines of research in the field of placebo studies. Each of them illuminates one particular question and together they demonstrate the rich variety of topics that have been addressed in placebo research during the past 40 years.

1. The power of mind (Reeves et al. 2007)

A 26-year-old man, Mr. A, presented himself to an emergency department after taking 29 capsules of a study drug. He had been depressed for about two months after his girlfriend broke up with him. After seeing an advertisement for a clinical trial of a new antidepressant he decided to enrol in the trial. Originally, A had felt that his mood had improved with this new medication. At the beginning of the second month he had an argument with his girlfriend, after which he decided to take all the remaining capsules. However, he felt immediately that it was a mistake and he asked his neighbour to take him to the hospital.

At the emergency department A was pale and tremulous. He was sweating and breathing rapidly and his blood pressure was 80/40. He was given two litres of saline, after which his blood pressure rose but it dropped again after the infusion was slowed. After a few hours it was found out that A had been in the control arm of the clinical trial and that he actually taken 29 capsules of placebos.

When A was informed about this, he felt surprised and relieved. Shortly thereafter he was alert and his blood pressure was normal again. He was admitted to the psychiatric unit from which he was later discharged diagnosed with depressive disorder and dependent personality disorder. He responded well to treatment with antidepressant medication and psychotherapy.

The patient had taken placebos but had experienced a *nocebo effect*, the mean twin brother of the placebo effect.

2. But does placebo effect exist, after all? (Hróbjartsson and Gøtzsche 2001)

A classical paper in the field of placebo research is Henry Beecher's The powerful placebo from 1955. The main message of that paper was that "placebos have a high degree of therapeutic effectiveness in treating subjective responses" (Beecher 1955). This conclusion has been cited in numerous other papers and in medical education for decades and it was questioned only in 2001 when Hróbjartsson and Gøtzsche published their–now also classical–meta-analysis "Is the placebo powerless?".

The approach of the meta-analysis was new: the researchers combined the results of 114 studies that included both a placebo arm and a no-treatment arm. Since the interest was in studying the possible placebo effect, the actual treatment arm was of no interest.

The results of the study provoked widespread interest around the world. A common conclusion both in the medical world and in the popular media was that a placebo effect did not exist at all. This conclusion was, however, misleading.

When compared with no treatment, placebos had no significant effect on subjective or objective *binary* outcomes (i.e. the patient had or did not have the outcome). If the outcome was *continuous*, a small but statistically significant effect was seen in the placebo group. In the trials involving the treatment of pain, the mean reduction in the intensity of pain was 6.5 mm on a 100 mm visual analogy scale.

Hróbjartsson and Gøtzsche pointed out that they did not study the possible effect of the patient-provider relationship. They also calculated that, disregarding possible biases, the pooled effect of a placebo on pain corresponds to one third of the effect of a non-steroidal anti-inflammatory drug. The authors' own conclusion was that there was little evidence that placebos in general have powerful clinical effects and that the use of placebos outside a clinical trial cannot be recommended.

The critics of the study pointed out that it was a good example of putting together "apples and oranges", i.e. performing a meta-analysis of studies that addressed very different issues, like the treatment of asthma, hypertension, anxiety, insomnia and tobacco addiction.

3. Yes, it does exist, even without placebos (Amanzio et al. 2001)

Use of placebos and placebo effects are often associated, as if the former would be a necessary condition for the latter. This is not the case, which was elegantly shown in a study published in the same year as the meta-analysis discussed above.

In the first part of the study the researchers used a so-called *open-hidden paradigm* to find out the role of 'non-specific' factors in pain relief after analgesics. Non-specific is a problematic but common term referring to other factors in the clinical setting than the 'specific' drug or method used.

Altogether 278 patients participated in the study. They had undergone thoracic surgery for different conditions and they all received the usual analgesic therapy postoperatively. The only difference between the patient groups was the way the drugs were administered. The *open* injections were given by a doctor in full view of the patients, who were told that they would receive a powerful painkiller. The *hidden* injections were administered by a pre-programmed infusion machine without any doctor or nurse in the room. The patients were thus unaware that a painkilling medication was being given.

The results were interesting: whatever the painkiller in question (buprenorphine, tramadol, ketorolac and metamizol), the hidden injections were significantly less effective and less variable compared with open injections.

According to the researchers, the placebo effect could be completely eliminated by means of hidden infusions. The difference in effect between the *hidden* and *open* groups could be explained by placebo effect, which in this case was caused at least by the presence of the physician, his or her words, gestures etc.

The *open-hidden paradigm* was used also in the second part of the study, in which experimental ischemic arm pain was induced in healthy volunteers. Also in this setting the response to a hidden injection of a painkiller (ketorolac) was smaller than to an open injection. No placebos were used in this study, yet an obvious placebo effect could be clearly demonstrated.

4. *It can be reversed with a drug* (Levine et al. 1978)

A landmark study on the neurobiology of the placebo effect was performed in 1978 in the University of California. The hypothesis was that placebo analgesia is mediated by endorphins.

The patients were young adults who had their wisdom teeth (mandibular third molars) removed. They all had diazepam before the operation and standard local anaesthesia during the operation. Three and four hours after the operation they were given an intravenous dose of either naloxone (a substance that blocks the effects of opioids) or placebo under randomised, double-blind conditions. Patients who received placebos first were classified in the analyses as placebo *responders*, whose pain was reduced or unchanged, or as *nonresponders* whose pain increased.

In general, patients given naloxone reported significantly greater pain than those given the placebo. In the nonresponders group naloxone given as a second drug produced no additional increase in pain levels. In the responders group, however, it *increased* the pain levels.

The final mean pain rating in the nonresponders group was identical to that in the responders group who had received naloxone as their second drug. It could be concluded that the enhancement of pain produced by naloxone was fully accounted for by its effect on placebo responders.

The study was small and it did not include a non-treatment group. In spite of these limitations it had shown for the first time that the placebo analgesia could be reversed pharmacologically.

5. *And you can see it in the brain* (Petrovic et al. 2002)

Since the ground-breaking study of Levine et al. (1978) numerous pharmacological and neuroimaging studies have demonstrated the biological mechanisms in placebo effects. Another landmark study was published in 2002, when Petrovic et al. demonstrated that both opioid and placebo analgesia are associated with increased activity in a brain area called the rostral anterior cingulate cortex (rACC).

The researchers compared the analgesic effects of remifentanil, a rapidly acting opioid and placebo treatment, using six different conditions: (1) heat pain and opioid treatment, (2) non-painful warm stimulation and opioid treatment, (3) heat pain and placebo treatment, (4) non-painful warm stimulation and placebo treatment, (5) heat pain only and (6) non-painful warm stimulation only (Petrovic et al. 2002).

In addition to behavioural responses the researchers used positron emission tomography (PET) to measure regional cerebral blood flow (rCBF) in certain brain areas. The aim was to compare the functional anatomy of the placebo analgesic response with that of the opioid response.

The main result of the study was that that *both* opioid and placebo analgesia are associated with increased activity in the rACC. There was also a correlation between the activity in the rACC and the brainstem during both opioid and placebo analgesia, but not during the pain-only condition. Together, these findings indicate a related neural mechanism in placebo and opioid analgesia (Petrovic et al. 2002).

6. *Branding matters* (Branthwaite and Cooper 1981)

It is commonly known that *branding* has an effect on our evaluation of products. A brand refers to the "name, term, design, symbol, or any other feature that identifies one seller's good or service as distinct from those of other sellers" (American Marketing Association 1995). The effect of branding of a well-known analgesic was addressed in another classical study of the treatment of headache in 835 women (Branthwaite and Cooper 1981). The study subjects were volunteers who had a history of using painkillers to relieve headache.

They were divided into four different groups. Group A was given placebos labelled "analgesic tablets" and group B placebos labelled with the name of a popular aspirin-based analgesic. Group C received aspirin tablets labelled "analgesic tablets" and group D aspirin tablets labelled with the same brand name as in group B. The packages were thus identical in groups A and C ("analgesic tablets") and in B and D (the brand name). Each subject was given a 50-tablet canister of the corresponding analgesic/placebo. They were instructed to take two tablets for any headache they had over the following two weeks.

It was found that the pharmacologically active formulations (aspirin groups C and D) worked better than placebos (groups A and B). Similarly, the branded preparations (B and D) worked better than the unbranded preparations (A and C).

In conclusion, branding of the preparation appeared to supplement the pain-relieving effect *both* in the placebo group *and* the aspirin group.

7. *Physicians' expectations matter, too* (Gracely et al. 1985)

Patients' expectations are often mentioned as a psychological mechanism that explains placebo effects at least partly. It is less common to discuss the expectations of clinicians, although a small study demonstrated their importance over 30 years ago.

Gracely et al. (1985) wanted to find out whether the clinician's knowledge of the range of possible treatments could influence the magnitude of the placebo effect in a double-blind study. They recruited 60 dental patients who all had a similar minor operation and gave their consent to participate in the study. The patients were told that they might receive saline (as a placebo in this study), fentanyl (a narcotic analgesic) or naloxone (a narcotic antagonist). The drugs and the saline were administered double blind. The patients assessed their pain with a questionnaire 10 min before the operation and 10 and 60 min after the operation.

The point of the whole study was that the clinicians were told a different story. According to the information given to them, one group (PN) would receive either placebo or naloxone and the other (PNF) group placebo, naloxone or fentanyl. All drugs were given double blind but the clinician knew which group the patients belonged to. They knew that the patients in group PN would receive either a placebo or naloxone that would *enhance* the pain.

The study was reported as a short letter in the Lancet and the only result presented was the change in the pain rating index in the patients who received the placebo. The difference between groups PNF and PN was dramatic: pain after placebo administration group *decreased* in the former but *increased* in the latter group. The sample size was very small but still the difference was statistically significant.

The only difference between the two groups was in the clinicians' knowledge of the range of possible treatments. In the case of group PN, the clinician knew that the patient could receive only placebo or a drug that enhances the pain, and in the case of group PNF, she knew that the patient could also receive a powerful painkiller. This knowledge somehow resulted in subtle behaviours influencing the patients.

The message of this small study is clear: it matters what the clinician expects. The finding is important and interesting and it has been cited numerous times but there is a neglected issue related to the ethics of the study. According to the report, the patients signed a consent form approved by an ethics committee at the National Institutes of Health (in the U.S.A). It is, however, obvious that either the patients or the clinicians were misled about the study drugs.

8. Placebos work even without deception (Park and Covi 1965)

It is still widely believed that response to placebos requires deception. This assumption was, however, questioned already in 1965, when two American doctors performed the first so-called open-label placebo study. They gave placebos to fifteen anxious and neurotic outpatients who were told that the pills contained inert material. Fourteen patients completed the study and all reported improvement. There was also marked improvement by doctor ratings on several measure. Eight patients believed that they had received placebos and six patients thought that the pills contained drugs. Interestingly, improvement was not related to belief in the nature of the pills but to certainty of the belief.

The study by Park and Covi was largely ignored or forgotten for decades until a new line of research on open-label placebos was established 40 years later.

9. Can placebo replace oxygen? (Benedetti et al. 2018)

On one hand, research during the past decades has demonstrated the wide variety of conditions in which placebo effects take place. On the other hand, there are conditions, in which no placebo is expected. It is unlikely, for example, that placebo antibiotics would kill bacteria or placebo anaesthetic gas would induce general anaesthesia. It is widely assumed also that placebo effects have no role in supporting critical life functions.

Recently, however, an Italian research group lead by Professor Fabrizio Benedetti has opened a brand new line of research by studying the question "can oxygen be partially replaced by a placebo?". According to the preliminary studies with a small number research subjects tha amazing answer is "yes". The researchers performed the experiments at high altitudes of 3500, 4500 and 5000 metres, where oxygen pressure is 64, 57 and 50%, respectively, compared to the sea level. Healthy volunteers completed exercise tests and the effects of placebo given for the first time, of oxygen, and of placebo after oxygen preconditioning on various symptoms and physiological parameters were measured. Marked placebo effects were found in overall performance and functions like ventilation and circulation but not in oxygenation. The researchers concluded that because these placebo effects take place without any change in oxygen concentration, psychobiological mechanisms can sometimes be more important than or as important as the oxygen content of the body.

10. Beyond medicine (Phillips et al. 1993)

Placebo effects have been described in the context of medicine, but it is obvious that medical placebo effects are part of a larger group of phenomena in human behaviour, communication and culture.

In a very large "natural experiment" the causes of death of 28,169 adult Chinese-Americans and 412,632 randomly selected, matched 'white' controls were examined. Altogether 15 leading causes of death were identified and two major causes of death—heart disease and cancer—were examined in more detail. The hypothesis was that culture and beliefs have an impact on morbidity and mortality. The core belief in the traditional Chinese culture is that a person's fate is influenced by the year of birth. For each Chinese-American person, 20 controls with the same sex, year of death and cause of death were selected.

It was found that "Chinese-Americans die significantly younger if they have a combination of disease and birth-year which Chinese astrology considers ill-fated, and that more years are lost by groups attached to Chinese traditions." The strength of attachment to Chinese traditions correlated to years lost in those with ill-fated combinations of disease and birth-year. In addition, females with ill-fated combinations lost more years than males, perhaps because they were less exposed to Western influences.

In the discussion section of their paper, Phillips et al. examine possible confounding factors, such as a change in the behaviour of the Chinese-American patients, their doctors, or death registrars. Their conclusion is that the findings result at least partly from psychosomatic processes, which cannot be precisely identified as yet. Anthropologist Daniel Moerman (2002) says it more clearly: these significant differences in longevity among Chinese-Americans are not "due to having Chinese genes, but to having Chinese ideas, to knowing the world in Chinese ways".

These ten research papers paint a colourful picture of the phenomena behind so-called placebo effects. In fact, they also demonstrate the inability of the term 'placebo effect' to cover all these phenomena and the need for a more precise terminology. But let us leave science for a while and look in another direction, namely literature.

1.2 Placebo Effects in Literature

Art and literature have the power to dramatize life in ways that the authors of scientific papers can only dream of. Various forms of placebo effects have also been described in novels and film, but less frequently than negative effects such as insanity, coma or death, provoked by shamans and witchdoctors (Marshall 2004).

While the term 'placebo effect' is uncommon in fiction, the related phenomena can be found in world literature. The mere presence of a physician (or a healer), the ritual she performs and the context as such may have a powerful impact on the condition of the patient. This will now be exemplified by eight extracts from literature (Louhiala and Puustinen 2013).

Plato: Charmides

Alfred North Whitehead famously wrote that the European philosophical tradition consists of a series of footnotes to Plato. The placebo effect as a phrase did not exist in Plato's time but the idea can, in fact, be found in *Charmides*, where the cure for headache is discussed in passing. Socrates was asked whether he knew the cure for headache. He replied that

> it was a kind of leaf, which required to be accompanied by a charm, and if a person would repeat the charm at the same time that he used the cure, he would be made whole; but that without the charm the leaf would be of no avail. (Plato 1980)

In today's medicine we may have more powerful 'leaves' and the 'charms' are different but the healing relationship and the meaning of the rituals are as important as they were in antiquity.

Tolstoy: War and Peace

Natasha Rostova, a central character in *War and Peace* by Leo Tolstoy (1828–1910), fell seriously ill after her engagement was broken:

> She could not eat or sleep, grew visibly thinner, coughed and, as the doctors made them feel, was in danger. They could not think of anything but how to help her. Doctors came to see her singly and in consultation, talked much in French, German, and Latin, blamed one another, and prescribed a great variety of medicines for all the diseases known to them, but the simple idea never occurred to any of them that they could not know the disease Natasha was suffering from, as no disease suffered by a live man can be known, for every living person has his own peculiarities and always has his own peculiar, personal, novel, complicated disease, unknown to medicine – not a disease of the lungs, liver, skin, heart, nerves, and so on mentioned in medical books, but a disease consisting of one of the innumerable combinations of the maladies of those organs.

The doctors made "the patient swallow substances for the most part harmful (the harm was scarcely perceptible, as they were given in small doses)." They assured Natasha

> that it would soon pass if only the coachman went to the chemist's in the Arbat and got a powder and some pills in a pretty box for a ruble and seventy kopeks, and if she took those powders in boiled water at intervals of precisely two hours, neither more nor less.

The usefulness of the physicians did not, however, depend on these medicines. They were useful and necessary, because

> they satisfied a mental need of the invalid and of those who loved her. … They satisfied that eternal human need for hope of relief, for sympathy, and that something should be done, which is felt by those who are suffering. They satisfied the need seen in its most elementary form in a child, when it wants to have a place rubbed that has been hurt.

Chekhov: A Case History

Anton Chekhov's (1860–1904) short story *A Case History* depicts a house call by a young physician, Dr. Koryolov, to a wealthy family living in the countryside. The patient was

> a girl of twenty – Liza, Mrs. Lyalikov's only daughter and heiress – who was ill. She had been unwell for some time, she had been under various doctors, and during the entire previous night she had suffered heart palpitations so acute that no one in the house had slept – they had feared for her life.

Dr. Koryolov performed a physical examination and found no organic problems. The treatment was reassurance and advice to get some sleep, but then something happened to change the situation:

> Then a lamp was brought into the bedroom. The sick girl squinted in the light, suddenly clutched her head in her hands—and burst out sobbing. … He saw a gentle, suffering look so wise, so moving that she seemed all feminine grace and charm – he wanted to soothe her, now, with a few simple kind words: not with medicines or advice.

As Koryolov was leaving the house, Mrs. Lyalikov asked him to, instead, to stay until the following day, and he agreed. When he came back to the house he heard the patient weeping. He entered her room and asked how she was.

> 'All right, thank you.' He felt her pulse, pushed back the hair which had fallen on her forehead. 'You can't sleep,' he said. 'There's wonderful weather outside – spring, nightingale song – but you sit brooding in the darkness.' She listened, gazed into his face. Her eyes looked sad and wise, she obviously wanted to tell him something.

With these small gestures, Koryolov formed a connection with his patient. A little later he interpreted her problem as a struggle with the questions of right and wrong. The short story has an open ending: the doctor leaves the house and the patient is seen standing on the porch, smiling sadly, as if hiding her innermost secrets.

Waltari: The Egyptian

The Egyptian is the first and most famous of the historical novels by Mika Waltari (1908–1979), a Finnish author known for the accuracy of the historical facts in his works (Saloheimo 2012). The book depicts the life of Sinuhe, a fictional physician working in ancient Egypt. He tells his story in exile after Pharaoh Akhenaten's fall and death in the 14th century B.C.

Sinuhe's father was Senmut, a poor but widely respected physician in his community. As a boy, Sinuhe used to help his father in his office. One day the wife of a spice dealer "wearing jewelry and a collar sparkling with precious stones" came for examination. She "sighed and moaned and lamented over her many afflictions" while Senmut listened carefully. Sinuhe describes the event:

> He wrote a line in ancient characters copied from a worn papyrus roll, then poured oil and wine into a mixing bowl and soaked the paper in it until the ink had been dissolved by the wine. Then he poured the liquid into the earthenware jar and gave it to the spice dealer's wife as a medicine, telling her to take some of it whenever head or stomach began to pain her. When the woman had gone, I looked at my father who seemed embarrassed. He coughed once or twice and said, 'Many diseases can be cured with ink that has been used for powerful invocation.' He said no more aloud, but muttered to himself after a time, 'At least it can do the patient no harm.'

Canetti: The Tongue Set Free

Elias Canetti (1905–1994), the Nobel laureate for literature in 1981, was a cosmopolitan "with one native land, the German language." He wrote in several genres but his most widely read work is his memoirs in three volumes, *The Tongue Set Free* (1977), *The Torch in the Ear* (1980), and *The Play of the Eyes* (1985).

In *The Tongue Set Free*, Canetti describes an episode from his childhood, when he was playing with his friend Laurica. The children were running inside the house, Laurica caught Elias close to a cauldron of hot water and he fell into it. He suffered from severe burns, and the family feared for his life. Elias' father was in England at the time, and his absence was even worse for the boy than the pain of the burns. He missed his father desperately and kept asking, "When is he coming?" The father left for home immediately when he heard of the accident, but, at that time, it took several days to travel across Europe. Canetti describes his father's arrival:

> Then I heard his voice, he came to me from behind, I was lying on my belly, he softly called my name, he walked around the bed, I saw him, he lightly put his hand on my hair, it was father, and I had no pains.
>
> Everything that happened from then on I know only from what I was told. The wound became a wonder, the recovery began, he promised not to go away anymore and he stayed during the next few weeks. The doctor was positive I would have died if my father hadn't come and remained. The doctor had already given me up but insisted on my father's return, his only, not very sure hope.

O'Brian: The Hundred Days

Patrick O'Brian (1914–2000) is best known for his 20 sea novels set in the Royal Navy during the Napoleonic Wars. One of the central characters was naval physician Stephen Maturin, the "ship's gifted surgeon, but he is also a scientist, an espionage agent for the Admiralty, a man of part Irish and part Catalan birth—and a revolutionary." (Hitchens 2003). Dr. Maturin had a salutary effect on his fellow mariners by his mere presence. Even after twenty years at sea he was a landsman and "nautical terms and naval routines seem to disappear through a mental sieve in his brainpan" (Marshall 2004). In many senses he was the untidy opposite of his naval colleagues, but

This general view, however, in no way affected their deep respect for him as a medical man: his work with a trephine or a saw, sometimes carried out on open deck for the sake of the light, excited universal admiration, and it was said that if he chose, and if the tide were still making, he could save you although you were already three parts dead and mouldy. Furthermore, a small half of one of his boluses would blow the backside off a bullock. The placebo effect of this reputation had indeed preserved many a shattered sailor, and he was much caressed aboard. (O'Brian 1998)

Lewis: Arrowsmith

The main character in Sinclair Lewis' novel *Arrowsmith* is Martin Arrowsmith, who already as a young boy spent his free time with Doc Vickerson, an alcoholic doctor in a small town. Arrowsmith went to college and medical school, began his career as the only doctor in a remote community and worked in various positions in hospital medicine and public health. Early in his career he was also interested in research:

Martin banged on the table and quoted his idol: "Up to the present, even in the work of Erlich, most research has been largely a matter of trial and error, the empirical method, which is quite the opposite of the scientific method, by which one seeks to establish a general law governing a group of phenomena so that he may predict what will happen. (Lewis 2008)

After publishing a scientific paper Arrowsmith was invited to work in a famous research institute. Later he was able to combine clinical work and research when he travelled to St. Huber, a remote Caribbean island endemic with the plague.

One of the most powerful moments in the book occurs when Dr. Arrowsmith tried "to convince the colonial government of St. Hubert to let him conduct a controlled clinical trial of his anti-plague serum, his 'phage'." (Marshall 2004, 39) The proposed trial did not contain a placebo arm but involved a randomized treatment methodology. *Arrowsmith* was published in 1925 and the idea of a controlled trial was not widely known at that time. However, Dr. Arrowsmith proposed that half of the island's citizens would receive the serum, while the other half of the islanders would receive treatment as usual:

He sought to explain that he could—perhaps—save half of a given district, but that to test for all time the value of phage, the other half must be left without it…though, he craftily told them, in any case the luckless half would receive as much care as at present. (Lewis 1925)

The response from the "Special Board" members was a moral outrage "bordering on apoplexy at the idea of withholding a potentially effective treatment from anyone." (Marshall 2004, 39). Arrowsmith was, however, finally allowed to pursue his study. Placebos were not involved but the scientific principle of a randomised controlled trial was finally accepted.

Rowling: Harry Potter and the Half-Blood Prince

In J.K. Rowling's *Harry Potter and the Half-Blood Prince* (2005) Ron Weasley, one of the central characters of the series, is picked to act as Keeper for the Gryffindor Quidditch team. To boost Ron's confidence, Harry pretends to give him *Felix Felicis*, also known as *Liquid Luck*, a potion which makes the drinker unusually lucky. Ron believes that he has actually taken it, plays very well and helps Gryffindor win the match. The main female character Hermione is angry with Harry:

> 'You know perfectly well what we're talking about!' said Hermione shrilly. 'You spiked Ron's juice with lucky potion at breakfast! Felix Felicis!'
>
> 'No I didn't,' said Harry, turning back to face them both.
>
> 'Yes you did, Harry, and that's why everything went right, there were Slytherin players missing and Ron saved everything!'
>
> 'I didn't put it in!' said Harry, now grinning broadly. He slipped his hand inside his jacket pocket and drew out the tiny bottle that Hermione had seen in his hand that morning. It was full of golden potion and the cork was still tightly sealed with wax. 'I wanted Ron to think I'd done it, so I faked it when I knew you were looking.' He looked at Ron. 'You saved everything because you felt lucky. You did it all yourself.'

All of these stories describe effects that are real and meaningful for the persons involved. Yet in today's medical terminology these effects would easily be labelled placebo effects, suggesting that the patients did not receive real or specific treatment. They accordingly illustrate why the term is unhelpful and misleading in the context of clinical reality. Moreover, while the term placebo refers to inert drugs or treatments, no such inert substances or fake treatments were used in any of the preceding stories.

It is hard to tell what Plato and Socrates really thought about the healing mechanism of the leaves accompanied by a charm, but the description fits perfectly to the understanding of a doctor-patient relationship today.

Dr. Koryolov, in *A Case History*, did not provide drugs, but only his words and his presence, and Dr. Maturin in *The Hundred Days* was praised for the placebo effect of his reputation. In *The Tongue Set Free*, the key person provoking the healing effect was not a physician, but rather the little boy's father. According to today's standards, the powder and pills given to Natasha in a pretty box in *War and Peace* were probably more harmful than helpful as such, and the ink dissolved in wine in *The Egyptian* was by no means biologically inert, although Waltari hints that Senmut did not actually believe the solution as such would be helpful.

The study setting described in *Arrowsmith* is an early illustration of a controlled trial, which became a golden standard of clinical research some 30 years later. Placebos are not mentioned since they were not part of the study routine in the 1920's. A modern version of the trial would include a comparison group receiving a placebo serum.

The episode described in *Harry Potter and the Half-Blood Prince* is a lively illustration of the placebo effect, or, in fact, one of the many placebo effects, as they are understood in today's medicine.

Like the ten scientific papers in Sect. 1.2, these examples from world literature demonstrate the large variety of phenomena behind so-called placebo effects. They show also that people have understood the existence and meaning of these phenomena long before 'placebo' or 'placebo effect' entered our vocabulary. Let us now take a first look at the conceptual confusion.

1.3   Conceptual Confusion—An Introduction

As Ben Goldacre writes in *Bad Science* (2011, p. xi), the placebo effect is "prob-ably the most fascinating and misunderstood aspect of human healing." However, confusion around the concepts prevails and often makes things look more complex than they really are. The confusion is common even among researchers of placebo effects, and they are—perhaps painfully—aware of the problem:

> … the medical and psychiatric literature on placebos and their effects is conceptually bewildering, to the point of being a veritable Tower of Babel. (Grünbaum 1986)

> Generally the conceptual and methodological confusion in the field of placebo is of such a magnitude that references to placebo effects are incomprehensible without further clarifica-tion. It might be time to stop using the term placebo effect and instead specify which kind of intervention one is referring to, and how its effect was measured. (Hróbjartsson 2002)

> A number of prominent scholars (drawn from medical practice, as well as philosophy, psy-chology, and anthropology) continue to propose and defend different conceptual models for these terms, and the perception that conceptual debate persists is often given as one justification for new definitions. (Blease 2018)

The main purpose of this book is to make sense of the concepts and the phenomena behind them. In the beginning of this chapter I have already mentioned some central misunderstandings but now it is time to explore them a little more. If you get confused, don't worry, you are not alone. All of the issues will be clarified later in this book.

What are placebos?

For most people a placebo refers simply to an *inactive substance,* used in medi-cal research or practice. In the placebo effect research context, however, the issue becomes more complicated. In a Danish study addressing the use of placebos in clinical practice a placebo treatment was characterized as "an intervention not con-sidered to have any 'specific' effect on the condition treated, but with a possible 'unspecific' effect." (Hróbjartsson and Norup 2003). The researchers asked, for example, how often during the last year the respondents had used "saline injections, B vitamins, antibiotics, sedatives, or physiotherapy in clinical situations in which the expected effect of the pharmacological or "specific" content of the interventions was negligible." (ibid.).

Sometimes the concept of placebo is understood very broadly, covering practically all aspects of the clinician-patient encounter, except the specific effect of a drug or other treatment modality. In their paper "Harnessing placebo effects in health care" Chaput de Saintonge and Herxheimer (1994) expand the realm of placebo "to the causes of the aggregated non-specific effects of treatments when specific effects have been segregated." A table in the paper classifies placebos into eight main classes: (1) scars; (2) pills, tablets, and injections; (3) appliances; (4) touch; (5) words; (6) gestures; (7) local ambience; and (8) social interventions.

Are placebos inert or not?

According to the standard definition of placebos they are inert or inactive. This sounds simple and easy but it is not. From a strictly biological point of view *no substances* in nature are fully inert. The placebos used in clinical research are, in most cases, inert enough, but sometimes even this is not obvious. Corn oil and olive oil capsules have been used as placebos in studies for cholesterol lowering drugs and lactose pills as placebos in a study on gastrointestinal symptoms in cancer patients. Since cancer patients bear an increased likelihood of lactose intolerance, the true nature of 'placebo' in this case remains uncertain (Golomb 2011).

The activeness or inactiveness of placebos is, of course, related to the previous question of our understanding of the concept. Benedetti (2011) relates the symbolic nature of placebos to their essence: "Placebos are not inert substances, as thus far believed … They are made of words and rituals, symbols, and meanings, and all these elements are active in shaping the patient's brain."

Are placebos used in medicine?

According to a textbook by two famous placebo researchers, "the knowing prescription of placebos has not been used in medicine for many years" (Shapiro and Shapiro 1997). Twenty years later two other well-known placebo researchers wrote, however, that "the placebo is arguably the most commonly prescribed drug, across cultures and throughout history" (Gold and Lichtenberg 2014). The latter claim has been repeated in several studies across the world and raises the following immediate questions: (1) Do physicians really use inert substances deliberately and frequently without telling their patients? (2) If physicians report that they use placebos, but they do not de facto refer to inert substances, what exactly do they mean by placebos?

Should placebos be used in clinical medicine?

The conceptual confusion around the concept placebo and placebo effect is reflected in the opinions about the use of placebos in clinical medicine. Quite commonly, deliberate use of inert substances is considered unethical (AMA 2018; Hróbjartsson 2008). Some prominent scholars have, however, presented opposite views: the American psychiatrist Walter Brown argued in 1998 that, "As physicians, we should respect the benefits of placebos—their safety, effectiveness and low cost—and bring the full advantage of these benefits into our everyday practices" (Brown 1998). Lichtenberg et al. (2004) went even further in 2004 when they asserted that "in select cases, use of placebo may even be morally imperative". Brewer (2012) even complained that "it's a pity that placebo prescribing is now almost impossible for proper doctors."

Are placebos needed for placebo effects?

Although the concepts placebo and placebo effect often go together, there is no necessary connection between them. Also, the worlds of clinical research and clinical practice are separate and generalisations should not be made between them. Studies using the *open-hidden paradigm* described above are perhaps the most powerful demonstration of the fact that placebos (understood as inert substances) are not needed to provoke placebo effects.

These five questions are different by nature. Some of them are purely empirical but some are conceptual or philosophical. The rest of this book is an attempt to clarify the conceptual issues and find answers to the empirical questions. At this point it should be clear that the conceptual problems are real and serious. Let us take a deeper loot at them.

References

AMA (American Medical Association). Use in clinical practice. https://www.ama-assn.org/delivering-care/use-placebo-clinical-practice. Accessed 1 Jun 2018.

Amanzio, M., A. Pollo, G. Maggi, and F. Benedetti. 2001. Response variability to analgesics: A role for non-specific activation of endogenous opioids. *Pain* 90: 205–215.

American Marketing Association Dictionary. 1995. https://www.ama.org/resources/Pages/Dictionary.aspx?dLetter=B. Accessed 1 Jun 2018.

Barrett, B., D. Muller, D. Rakel, D. Rabago, L. Marchand, and J.C. Scheder. 2006. Placebo, meaning and health. *Perspectives in Biology and Medicine* 49: 178–198.

Beecher, H.K. 1955. The powerful placebo. *JAMA* 159: 1602–1606.

Benedetti, F. 2013. Placebo and the new physiology of the doctor-patient relationship. *Physiological Reviews* 93: 1207–1246.

Benedetti, F., E. Carlino, and A. Pollo. 2011. How placebos change the patient's brain. *Neuropsychopharmacology Reviews* 36: 339–354.

Benedetti, F., D. Barbiani, and E. Camerone. 2018. Critical life functions: Can placebo replace oxygen? *International Review of Neurobiology* 138: 201–218.

Blease, C. 2018. Consensus in placebo studies—Lessons from the philosophy of science. *Perspectives in Biology and Medicine* 61: 412–429.

Branthwaite, A., and P. Cooper. 1981. Analgesic effects of branding in treatment of headaches. *BMJ (Clinical Research Ed)* 282: 1576–1578.

Brewer, C. 2012. The golden age of placebo. *BMJ* 344: e3016.

Brown, W.A. 1998. The placebo effect. *Scientific American* 278 (1): 90–95.

Canetti, Elias. 2011. *The tongue set free*. Croydon: Granta Books.

Chaput de Saintonge, D.M., and A. Herxheimer. 1994. Harnessing placebo effects in health care. *Lancet* 344: 995–998.

Chekhov, A. 1990. *The princess and other stories*, trans. and ed. Ronald Hingley, 179–188. Aylesbury: Oxford University Press.

Furukawa, T.A., A. Cipriani, L.Z. Atkinson, S. Leucht, Y. Ogawa, L. Takeshima, et al. 2016. Placebo response rates in antidepressant trials: A systematic review of published and unpublished double-blind randomised controlled studies. *Lancet Psychiatry* 3: 1059–1066.

Gold, A., and P. Lichtenberg. 2014. The moral case for the clinical placebo. *Journal of Medical Ethics* 40: 219–224.

Goldacre, B. 2011. *Bad science*. London: Fourth Estate.

Golomb, B.A. 2011. The Dece(i)bo Effect. The Edge 2011. https://www.edge.org/response-detail/11708. Accessed 1 Jun 2018.

Gracely, R.H., R. Dubner, W.D. Deeter, and P.J. Wolskee. 1985. Clinicians' expectations influence placebo analgesia. *Lancet* 325: 43.

Grünbaum, A. 1986. The placebo concept in medicine and psychiatry. *Psychological Medicine* 16 (1): 19–38.

Hitchens, C. 2003. Empire falls—How master and commander gets Patrick O'Brian wrong. Slate. November 14, Accessed 1 Jun 2018.

Hróbjartsson, A. 2002. What are the main methodological problems in the estimation of placebo effects? *Journal of Clinical Epidemiology* 55 (5): 430–435.

Hróbjartsson, A. 2008. Clinical placebo interventions are unethical, unnecessary and unprofessional. *Journal of Clinical Ethics* 19: 66–69.

Hróbjartsson, A., and P.C. Gøtzsche. 2001. Is the placebo powerless? an analysis of clinical trials comparing placebo with no treatment. *New England Journal of Medicine* 344: 1594–1602.

Hróbjartsson, A., and M. Norup. 2003. The use of placebo interventions in medical practice: A national questionnaire survey of Danish clinicians. *Evaluation and the Health Professions* 26: 153–165.

Kaptchuk, T. Interview in 2012. https://www.youtube.com/watch?v=2rt7WIK2OVE. Accessed 1 Jun 2018.

Kaptchuk T.J.F.E., J.M. Kelley, M.N. Sanchez, E. Kokkotou, J.P. Singer, M. Kowalczykowski, et al. 2010. Placebos without deception: A randomized controlled trial in irritable bowel syndrome. *PLoS One* 5: e15591.

Levine, J.D., N.C. Gordon, and H.L. Fields. 1978. The mechanism of placebo analgesia. *Lancet* 2: 654–657.

Lewis, S. 1925. *Arrowsmith*. New York: Harcourt, Brace & World Inc.

Lichtenberg, P., U. Heresco-Levy, and U. Nitzan. 2004. The ethics of the placebo in clinical practice. *Journal of Medical Ethics* 30: 551–554.

Louhiala, P., and R. Puustinen. 2013. Placebo effect or care effect? four examples from the literary world. Hektoen International, vol. 5, issue 4. http://hekint.org/2017/03/03/placebo-effect-or-care-effect-four-examples-from-the-literary-world/?highlight=louhiala. Accessed 1 Jun 2018.

Louhiala, Pekka. 2012. What do we really know about the deliberate use of placebos in clinical practice? *Journal of Medical Ethics* 38: 403–405.

Marshall, M.F. 2004. The placebo effect in popular culture. *Science and Engineering Ethics* 10: 37–42.

Moerman, D. 2002. *Meaning, Medicine and the 'Placebo Effect'*. Cambridge: Cambridge University Press.

Moerman, D. 2013. Against the "placebo effect": A personal point of view. *Complementary Therapies in Medicine* 21: 125–130.

Nayernouri, T. 2017. Homeopathy, ritual and magic. *Archives of Iranian Medicine* 20: 718–722.

O'Brian, P. 1998. *The hundred days*. New York: W.W. Norton & Company.

Park, L.C., and U. Covi. 1965. Nonblind placebo trial: An exploration of neurotic patients' responses to placebo when its inert content is disclosed. *Archives of General Psychiatry* 12: 36–45.

Petrovic, P., E. Kalso, K.M. Petersson, and M. Ingvar. 2002. Placebo and opioid analgesia—Imaging a shared neuronal network. *Science* 295: 1737–1740.

Phillips, D.P., T.E. Ruth, and L.M. Wagner. 1993. Psychology and survival. *Lancet* 342: 1142–1145.

Plato. 1980. The collected dialogues of plato. In *Charmides*, ed. Hamilton E, Cairns H. Princeton: Princeton University Press.

Puustinen, R., and P. Louhiala. 2013. The paradox of placebo—Real and sham in medicine. In *The medical humanities companion volume three: Treatment*, ed. P. Louhiala, I. Heath, and J. Saunders. Oxford: Radcliffe.

Reeves, R.R., M.E. Ladner, R.H. Hart, and R.S. Burke. 2007. Nocebo effects with antidepressant clinical drug trial placebos. *General Hospital Psychiatry* 29: 275–277.

Rowling, J.K. 2005. *Harry Potter and the Half-Blood Prince*. London: Bloomsbury.

Saloheimo, P. 2012. Medical classics: The Egyptian. *BMJ* 345: e4712.

Shapiro, A.K., and E. Shapiro. 1997. *The powerful placebo—From ancient priest to modern physician*. Baltimore: The Johns Hopkins University Press.

St. Augustine. 1838. *Confessions*. Oxford: Parker and Rivington.

Tolstoy, Leo. 2001. *War and peace,* trans. and ed. Louise and Aylmer, Maude. London: Wordsworth 1. Classics.

Vogt, H., E. Ulvestad, T.E. Eriksen, and L. Getz. 2014. Getting personal: Can systems medicine integrate scientific and humanistic conceptions of the patient? *Journal of Evaluation in Clinical Practice* 20: 942–952.

Waltari, M. 1983. *The Egyptian*. Porvoo: Werner Söderström Osakeyhtiö.

Chapter 2
The Concepts

Discussion about conceptual or definitional matters in science may appear to be esoterica, however definitions are important. How we understand placebo concepts carries subtle but significant methodological implications for clinical trials as well as for ethical practice in the delivery of care. Therefore, gaining clarity about the argumentation within disputes over concepts is not trivial—rather, it might even be viewed as a major priority for the field of placebo research. (Blease 2018)

The concepts of placebos and placebo effects refer to phenomena that are incredibly diverse. They have been classed together only because they have been defined in a negative and theory-relative way. (Alfano 2015a)

…how ironic it was that a gathering of world experts in placebo studies, scholars who had collectively written thousands of per-reviewed articles and scores of books on the topic, could not agree on a common definition of the placebo effect. (Kelley 2018)

Medical theory is built on medical concepts that differ from everyday concepts since the former acquire their meaning as a part of the theoretical structure of medicine. According to Puustinen (2011), "Medical concepts are in a constant state of change, emerging and disappearing from use. It follows that to understand the content of medical concepts … we need to examine their developmental history in relation to medical theory."

It has been argued that until the 20th century, "despite medical advances, the vast majority of treatments were ineffective or placebos" (Shapiro and Shapiro 1997). While it is probably true that the drugs or methods *as such* were ineffective, it is highly questionable whether they could be called placebos.

An apparent scientific disagreement may be explained simply by the fact that disagreeing authors have defined the concepts in different ways. As Andersen (2005) notes, "The idea that something can be a placebo or a placebo effect belongs to a specific historical period which began at the end of the 18th century and lasts until today." The past cannot by judged by concepts from the present.

© Springer Nature Switzerland AG 2020
P. Louhiala, *Placebo Effects: The Meaning of Care in Medicine*,
International Library of Ethics, Law, and the New Medicine 81,
https://doi.org/10.1007/978-3-030-27329-3_2

Another question is whether exact definitions are necessary in medicine. It may well be that we will never be able to agree on necessary and sufficient conditions for a placebo or placebo effect. Alfano (2015b) suggests that we should "treat 'placebo' (and 'placebo effect') as philosophers of language have treated 'jade'". Many might suppose jade to be a natural kind of mineral, but it is not because there are two distinct minerals called 'jade', jadeite or nephrite (Bird and Tobin 2017). Alfano (2015b) continues: "Just as natural scientists study jadeite and nephrite but not jade as such, so medical researchers and medical ethicists can study various mechanisms that tend to produce what have been called placebo effects but not the placebo effect as such."

In addition, it is not meaningful to think about 'a placebo' or 'a placebo effect' without a historical context. That is why I start by exploring the early history of the concepts.

2.1 Early History of the Concepts

Placebo

The word placebo is the first-person future indicative of the verb *placeo* (to please). The expression can be found in several ancient Latin texts, such as Petronius' *Satyricon*, Seneca's *De Consolatione* and Martial's *Epigrammata*, to name a few (see References for internet sources).

Placebo entered the English language in the ecclesiastical context in the 13th century (Edwards 2005). In the Medieval Catholic Church 'Singing the Placebo' referred to the Office of the Dead ritual, in which Psalm 116 (Psalm 114 in the Latin Bible) was recited during the funeral. In the *Vulgate*, the 4th century Latin translation of the Old Testament by St. Jerome, Psalm 114:9 reads "placebo Domino in regione vivorum" ("I shall please the Lord in the land of the living").

The Latin text was not, however, the correct translation of the original Hebrew sentence (*et'halekh liphnay adonai b'artzot hakhayim*) (Aronson 1999). In the English Bibles the equivalent verse is Psalm 116:9, and it is correctly translated for example in the King James Version as "I will walk before the Lord in the land of the living". According to Aronson (1999)," St. Jerome may have reasoned that anyone who walked before (or as the Revised English Bible has it "in the presence of") the Lord would please Him, and used the word placebo for the sake of euphony or metre."

In medieval France it was customary for the dead person's relatives to give food to the congregation after the Office of the Dead ritual. Some people, often unrelated to the family, attended the service and sang the Placebo, hoping to be rewarded. Obviously this custom was common, since these fake mourners came to be known in French as 'placebo singers'.

By the 14th century the expression 'placebo singer' had been well established in the English language, too. In Geoffrey Chaucer's *The Merchant's Tale* (2001), for example, one of the central characters is *Placebo*, a sycophantic flatterer.

It took another four centuries before the word placebo found its way into the medical literature. In 1763 Alexander Sutherland, a British physician, wrote about a local charlatan *Placebo* who "never saw a professor in his chair, nor never made up a Doctor's prescription. Without knowledge chemical or practical, he was said to understand the waters *better than them all*." (italics original) (Sutherland 1763).

A rather modern characterisation of placebo as an ineffective substance was given in 1770 by Andrew Duncan, a Scottish physician: "Where a *placebo* merely is wanted, the purpose may be answered by means, which, although perhaps reduced under the *materia medica*, do not, however, deserve the name of medicines. When a class of medicines, then, is said to be indifferent with regard to a morbid affection, nothing farther is meant, than that is has no peculiar tendency to increase the evil; while, at the same time, no peculiar benefit can be expected from its employment." (Italics original) (Duncan 1770)

The most widely known and cited 18th century physician who used the term placebo was, however, William Cullen (Kerr et al. 2008). He described the treatment of Mr. Gilchrist, who was incurably ill. While acknowledging that he could not cure the patient, Cullen decided to prescribe him placebo, perhaps in order to comfort the patient: "I prescribed therefore in pure placebo, but I make it a rule even in employing placebos to give what would have a tendency to be of use to the patient" (Cullen 1772).

Kerr et al. (2008) point out that Cullen's use of 'placebos' to please patients is clearly discontinuous with early medieval and religious uses of the term. His use of the term was neutral, in contrast with the previous pejorative meaning that referred to flattery and fraud (Kerr et al. 2008). It is also worth noting that in most cases 18th century physicians did not administer *inert substances* as placebos but "resorted to any kind of medicine which they thought simple, feeble, or altogether powerless, non-perturbing medicines" (Jutte 2013).

One of the ways to investigate the emergence of a term into scientific language is to follow the development of dictionaries. The entry of the term placebo into English medical dictionaries took place in the late 18th century. The first edition of *Motherby's New Medical Dictionary* from 1785, for example, did not mention placebo at all, but the third edition from 1791 defined placebo as "a common place method or medicine." (Motherby 1785; Motherby 1791). It is not obvious what was meant by this definition. Shapiro and Shapiro (1997) speculate that 'commonplace' in the definition referred to banal, superficial, ordinary, dull or a cliché, as in the dictionary definitions of the word.

Similarly, *Hooper's Medical Dictionary* from 1798 did not know the term placebo but the 1811 edition included the following entry: "Placebo. I will please: an epithet given to any medicine adapted more to please than benefit the patient." (Hooper 1798; Hooper 1811). The definition is identical with the one given in *Coxe's Philadelphia Medical Dictionary* in 1808 (Coxe 1808).

By the first decades of the 19th century the term placebo seems to have become a part of physicians' clinical vocabulary. An indirect indication of the common understanding of the term can be found also in fiction. An example is a discussion between a patient and a doctor in Sir Walter Scott's novel St Ronan's Well, from 1823. The doctor responds to his patient: "You mistake the matter entirely, my dear Mrs. Blower, …., there is nothing serious intended—a mere *placebo*—just a divertisement to cheer the spirits, and assist the effect of the waters—cheerfulness is a great promoter of health." (Scott 1824)

The original Latin verb *placeo* is, of course as neutral as its English translation *to please*. As we have seen, however, negative connotations were attached to the term placebo when it was adopted in medieval French and English. These pejorative meanings were further transferred to the medical definitions and use of the term. An example can be found in an Editorial of the Edinburgh medical and surgical journal from 1834: "…it would appear that the homoeopathic plan of treating diseases is totally inert, and can be useful only as a placebo to hypochondriacs and nervous women, by relieving them from swallowing the manifold drugs which they think is their duty to burden their stomachs." (Editorial 1834)

A quantitative analysis of the use of the term placebo in the BMJ during the 19th century paints a colourful picture (Raicek et al. 2012) and demonstrates that the understanding of the term was about the same as it is today among the medical profession. The authors found 71 citations containing the term and distinguished nine categories for the use of 'placebo'. The commonest (31%) use was pejorative and referred to "no effect". Other citations "25% portrayed placebo treatment as permitting the unfolding of the natural history (the normal waxing and waning of illness), 20% described placebo as important to satisfy patients, 10% described it as fulfilling a physician's performance role, 4% described its use to buy time, 4% described its use for financial gain, 3% used it in a manner similar to a placebo control, and only one implied that placebo could have a clinical effect." Only one case reported disclosure. Water, disguised as morphine, was injected "hypodermically" to a patient with morphine addiction. Three months later the patient looked very well and had "a good laugh [with the doctor] over the pious fraud of the water hypodermic injection." The treatment was obviously successful since the patient said he would never touch opium again (Stuart 1889).

Placebo effect

The term "placebo effect" was not common until the last century, but it also dates back to the 18th century, the time when placebo established its place in the medical vocabulary. In 1776 Scottish physician W. Robertson wrote in his book *Observationes miscellaneae inaugurales De vino praecipue* how "Ex modo et quantitate quibus administratur, nihil nisi placebo effe conludere volo". (The medication given had nothing but pleasing effect to his patient) (Robertson 1776).

It is difficult to trace the use of the term "placebo effect" in the 19th century but by the turn of the 20th century it seems to have been in common use. Pigg (1900) described in *The International Journal of Surgery* in 1900, how in the course of the assisted delivery his role was "to prohibit all pulling in the first stage, and dispense with chloroform entirely, unless it be a little for its placebo effect – not enough to arrest contractions." Oswald (1902) discussed indigenous healing methods among Africans and wrote that "there may have been a mere placebo effect about the procedure…where a victim of serpent bites was dosed with a decoction of boiled ants…" Twenty years later Graves (1920) published a case report of a 15-year-old boy with delayed puberty and epileptic seizures. Graves gave him testicular extract in tablet form and the boy's symptoms declined gradually. According to the author, this could, however, have been related to the "placebo effects of the drugs given prior to admission".

The term 'placebo effect' became widely known only in 1955 when Henry Beecher published his famous article "The powerful placebo" (Beecher 1955), which was both a review on the topic and a meta-analysis of 15 studies covering a wide variety of conditions. Beecher found a "relatively constant" therapeutic effectiveness of 35% for placebos and concluded that it suggested "a fundamental mechanism in common" for placebos. That 35% (or "approximately one third") became a standard reference for placebo effect both in subsequent trials and medicine in general.

Beecher's methodology and assumptions have been heavily criticized but this figure is still commonly cited for example in medical education. I will describe Beecher's study and the criticism of his work in more detail in Sect. 3.4.

2.2 Modern Definitions and Their Problems

As we have seen above, the terms placebo and placebo effect have been used in a number of inconsistent ways and the crucial difference between placebo use and placebo effect is often not understood (Louhiala 2009). Examples of this confusion can be found even in the texts of well-known placebo researchers. Consider the following:

> In the broader ethics literature, some commentators on informed consent and nondeceptive therapeutics caution against the use of placebos in medical practice. Others propose that the placebo effect can be harnessed in various therapeutic contexts that do not pose ethical dilemmas. (Sherman and Hickner 2008)

Although the authors seem to suggest as much, there is no necessary connection between the use of placebos (however they are understood) and placebo effects (however they are understood).

Because of the ambiguity of the concepts, many researchers have given up trying to find satisfactory definitions for them. Brody (2000) writes that "defining 'placebo' and placebo effect in a logically coherent fashion is difficult, perhaps impossible". Gøtzsche (1994) admits that "I have tried to define placebo in an unambiguous, logically consistent, and testable way, and I have failed". He finds some consolation in philosophy: "it gives me some comfort that this state of affairs is not unusual in philosophy. We cannot define what constitutes a chair but this fact does not prevent us constructing them".

Philosopher Charlotte Blease (2018) has, indeed, contributed to the discussion about definition along these lines. She suggests that "scientific concepts of placebo and placebo effect should be differentiated from those meanings ascribed to the terms in other non-expert domains including among medical practitioners, clinical investigators, patients, and research participants." (ibid., p. 2). She notes further that "even within scientific contexts, conceptual stability can be typified by implicit understanding rather than articulable, explicit definitions among many scientists." (ibid., p. 2).

I have described above the early history of the concepts that illustrates their evolution. The modern history of the concepts and their definitions, reveals, in turn, a lot about the phenomena we are trying to address with these concepts.

Placebo

In May 2018 The *Wiktionary* defined placebo as "a dummy medicine containing no active ingredients; an inert treatment" (https://en.wiktionary.org/wiki/placebo). This definition represents well the general current understanding of the concept placebo among both lay people and the medical profession.

At the same time *Merriam-Webster* dictionary gave two definitions, the first of which refers to an inert substance. The second one, however, broadens the scope: "a usually pharmacologically inert preparation prescribed more for the mental relief of the patient than for its actual effect on a disorder." (https://www.merriam-webster.com/dictionary/placebo). As Howick (2017) remarks, "this is only coherent if we presume a Cartesian distinction between mind and body, a view whose untenability every serious investigator accepts, yet which nonetheless continues to cloud much thought in this area." In addition, contrasting 'mental relief' and 'actual effect' is derogatory for anyone who has ever been in need of mental relief. Howick (2017) also points out that "importing into the definition the reasons why a treatment is given is a mistake: the intentions of a clinician are one thing, the objective facts about physical processes another (though one hopes that the two are at least sometimes linked)."

Among the scientific community broader and more complicated definitions for the concept placebo and its derivatives have been suggested. Often they blur the distinction between placebo as such and placebo effect and smuggle the context or the therapist into the definition of placebo. It is also worth noting that 'a placebo', 'the placebo' and 'placebo' may refer to different things, and that terms often include at least some aspects of 'placebo effect' also: "Although the term placebo initially referred to a biologically inert substance, it has been generalized to include sham surgeries, attention-placebo conditions in evaluating psychological interventions, and incidental features of the therapeutic context (often called common factors), such as the warmth or authoritativeness of the therapist, that may interact with outcome." (Bootzin and Bailey 2005)

Along the same line, Benedetti includes the context of treatment into the definition of placebo: "…a placebo would be better defined as an inert treatment plus the context that tells the patient a therapeutic act is being performed." (Benedetti 2009).

The philosopher of science Adolf Grünbaum (1986) made an ambitious attempt in the 1980's to define the placebo concept. His definition is very long and complicated and I will repeat it here. It does not work as a *general* definition of placebo because he does not make a difference between placebo use and placebo effect. On the very first page of his paper, before giving his own complex definition, he demonstrates this confusion by writing about "placebogenic phenomena" and "the medical and psychiatric literature on placebos and their effects".

Grünbaum's definition has been described as "arguably the most sophisticated model in the literature" (Alfano 2015a) and "by far the best proposal for a formal definition of placebo" (Hróbjartsson 2002). In spite of these words, neither Alfano nor Hróbjartsson find it satisfactory. According to Hróbjartsson (2002) it is not a good conceptual clarification, "partly because he does not give any clear criteria for what constitutes a good therapeutic theory, nor what to do in situations when two theories compete". Alfano (2015a) points out that the "definition is theory-relative to the extent that what counts as a placebo or a placebo effect depends on what the prescribing physician (or the community of medical experts, or the patient, or anyone else) thinks or espouses."

Analogously with a part of Grünbaum's definition, the American Medical Association brings the beliefs of the physician into the definition, when it defines placebo as "a substance provided to a patient that the physician believes has no specific pharmacological effect upon the condition being treated" (AMA 2018). The problem of this definition is obvious. A substance (or method) may be a placebo today but not tomorrow (or vice versa), depending on the beliefs of the physician. Equally, a substance (or method) is a placebo when given by Dr. A (who believes it to be ineffective) but not when given by Dr. B (who believes the contrary) (Louhiala and Puustinen 2017).

Another philosopher of science, Jeremy Howick (2017), has presented a new version of Grünbaum's definition, in which he gives credit to the original but introduces four modifications to it. He is specially interested in the research context and argues that, with these modifications, "Grünbaum's definition provides a defensible account of placebos for the purpose of constructing placebo controls within clinical trials" (Howick 2017). His analysis is plausible in that context but his suggestion for investigating the implications of this definition for the ethics of placebos in clinical practice ignores the fundamental difference between clinical research and clinical practice.

Some authors refer specifically to a 'clinical placebo'. Gold and Lichtenberg (2015), for example, define it as "a medical intervention that attempts to modify the patient's condition by reinforcing psychosocial healing forces for the benefit of the patient via top-down mechanisms." The definition is somewhat vague and 'clinical placebo' seems to include the therapeutic relationship and the setting of the encounter. The authors do not specify 'top-down' but refer obviously to a causal chain from the psychological to the physiological.

As we saw in Sect. 1.3., an extremely broad meaning for placebo is given by Chaput de Saintonge and Herxheimer (1994), who expand the realm of placebo "to the causes of the aggregated non-specific effects of treatments when specific effects have been segregated". This characterisation seems to cover practically all elements of therapeutic encounter, except a specific pharmacological or other physiological mechanism. The authors even "outline the steps needed in planning placebo treatment" and "suggest a framework of possible placebo components of treatment" (ibid.). With regard to placebo medications, for example, they suggest the following:

Choose treatments whose appearance or route of administration is known to be associated
with strong placebo effects. Larger capsules may have greater effects than smaller ones, and
injections are more effective than tablets. Yellow capsules have been reported to have stimulant or antidepressant effects, and white capsules analgesic or hypnotic actions. Reinforce
pharmacological effects by specific suggestion, e.g., the effect of a bronchodilator can be
doubled by specific 'bronchodilator' suggestions. Do not use placebos alone. (ibid.) [original
references omitted]

Chaput de Saintonge and Herxheimer do a good job, in fact, in reviewing empirical studies about the elements of doctor-patient relationship and the setting of the
encounter that have been shown to enhance patient welfare. What they fail to do,
however, is to justify why the concept of placebo is needed here. They also use
uncritically the terms specific and unspecific, which is very common in this context.

More recent derivatives of the term placebo entering into medical vocabulary are
'pure' and 'impure' placebo. The concepts were introduced originally in the 1940s
at a Cornell Conference on Therapy, the proceedings of which were published in
1947 (Gold et al. 1947). In that publication, DuBois divided placebos into three
classes: (1) pure placebos (e.g., bread pills or lactose tablets with no significant
physiological effects), (2) impure placebos ("adulterated with a drug that might have
some pharmacological action, such as tincture of gentian or a very small dose of
nux vomica"), and (3) "the universal pleasing element which accompanies every
prescription."

These concepts were hardly ever mentioned in medical literature for decades
after their introduction and even in the context of empirical placebo research they
have been used only recently (Howick et al. 2013). According to Howick et al.
(2013), "Pure placebos are interventions such as sugar pills … or saline injections
without direct pharmacologically active ingredients for the condition being treated.
Impure placebos are substances, interventions or 'therapeutic' methods which have
known pharmacological, clinical or physical value for some ailments but lack specific
therapeutic effects or value for the condition for which they have been prescribed."

Although used in the research context, pure and impure placebo remain highly
ambiguous. I will discuss them in more detail in Sect. 4.2.

Placebo effect

The ambiguity of the concept 'placebo' is often transferred into the definitions of
'placebo effect'. Stewart-Williams and Podd (2004), for example, define placebo
effect as "a genuine psychological or physiological effect, in a human or another
animal, which is attributable to receiving a substance or undergoing a procedure, but
is not due to the inherent powers of that substance or procedure". "Attributable to"
seems to refer to a causal connection but this is not further elaborated.

Another example of the ambiguity is seen in a textbook by Shapiro and Shapiro
(1997) who define placebo effect as "primarily the nonspecific psychological or
psychophysiological therapeutic effect produced by a placebo, but may be the effect
of spontaneous improvement attributed to the placebo".

Earlier they have defined placebo as "any treatment … that is used for its ameliora-
tive effect on a symptom or disease but that actually is ineffective or is not specifically

effective for the condition being treated." If this is inserted into their definition of placebo effect, we end up with the following: "… therapeutic effect produced by any treatment … that actually is ineffective or is not specifically effective for the condition being treated … may be the effect of spontaneous improvement attributed to the [treatment]." The first part makes no sense and the second part limits the therapeutic result to spontaneous improvement only and rules out all other components of the therapeutic encounter (Moerman and Jonas 2002; Puustinen and Louhiala 2013).

Simplistic and sometimes outright false claims are also common in definitions of placebo effect. Braillon (2009), for example, writes that "pragmatically, the placebo effect is simply a belief". According to Barnhart (2014), "a placebo effect is the tendency of any medication or treatment, even an inert or ineffective one, to exhibit results simply because the recipient believes that it will work". Both authors are correct in the sense that beliefs (or more precisely, expectations) are a central psychological mechanism of placebo effects. There are, however, important other mechanisms like classical conditioning. In addition, the beliefs of the care-giver are as important as the beliefs of the recipient. Braillon (2009) also ignores the crucial difference between the research context and the clinical context, when he writes: "the subjective patient-reported alleviation is small, observed in only one third of the subjects and only under certain conditions". He does not give a reference but apparently refers to Hróbjartsson and Gøtzsche (2001) and Beecher (1955). The former showed that the effect in the clinical trial context is small but they did not say anything about the clinical practice context. "One third" is a myth that was created by Beecher (1955) and that has been debunked thoroughly (e.g., Kienle and Kiene 1997).

Another false belief is that deception is necessary to evoke a placebo effect (Chan 2014). This is not the case in research nor in clinical practice. In clinical trials the subjects are correctly informed that they may end up in a placebo group. In medical practice it is wrong to lie about the nature of the drug or method but placebos are not even needed for the placebo effects.

Very often the concepts placebo effect and placebo response are used interchangeably but for some authors their distinction is crucial. Schedlowski et al. (2015), for example, define placebo effect as "the symptom improvement after inert treatments in clinical trials" which consists of factors like the natural history of a disease, fluctuation of symptoms, response biases, effects of simultaneous other interventions, or regression to the mean. For Schedlowski et al., a *subset* of placebo effect is placebo response which "refers to the outcome caused by a placebo manipulation". It "is mediated via three interdependent factors: patients' expectations about treatment benefits, the quality and quantity of doctor-patient communication, and associative (conditioning) learning processes". Schedlowski et al. also remark that "placebo responses are not restricted to placebo treatments—they can also modulate the outcome of any active treatment."

Ernst and Resch (1995) made, in fact, the same distinction 20 years earlier, but used different terminology. For them the 'true placebo effect' was the same as Schedlowski's 'placebo response', and 'perceived placebo effect' the same as his 'placebo effect'. According to Foddy (2009), "physicians can avoid using a placebo, yet produce a placebo-like effect through the skilful use of reassurance and encouragement". No wonder readers of the scientific literature are confused.

For Brody (2009), a placebo response is "a change in the patient's health or bodily state that is attributable to the symbolic impact of medical treatment or the treatment setting". Also, "a placebo response would be predicted whenever a conscious patient engages in any form of medical encounter (or self-treatment activity)". Here the author makes it clear that causal factor is not the (inert) treatment but the symbolic meaning of the treatment and/or the setting.

Along the same lines, Miller et al. (2009) define that "the 'placebo effect' is a generic name for beneficial effects that derive from the context of the clinical encounter, including the ritual of treatment and the clinician-patient relationship, as distinct from therapeutic benefits produced by the specific or characteristic pharmacological or physiological effects of medical interventions". They further characterize the placebo effect "as a form of interpersonal healing, as distinct from spontaneous natural healing and from technological healing dependent on physiologically active pharmaceuticals or procedures". Their final argument is that "research on the placebo effect has the potential to revitalize the art of medicine". I do not think the art of medicine has ever died but otherwise agree: both theoretical and empirical placebo research is highly relevant for the ways doctor-patient relationships and health care in general are organised. This topic is further examined in Chap. 5.

As noted earlier, more important than an explicit definition may be the implicit understanding of the phenomenon. Accordingly, Blease (2018) suggests that, within the empirical field, "placebo effects are understood to be positive health changes that occur as a result of specific psychobiological mechanisms ... These psychobiological mechanisms are elicited, in turn, by a range of cues in the context of the practitioner-patient encounter".

Confusing placebo use and placebo effects

Throughout the scientific literature it is common to use the terms in ways that confuse more than clarify the issues.

'A placebo' or 'the placebo' sometimes refers to *both* the use of placebos and the placebo effect. This seems to be the case when Bhugra et al. (2015) write "placebo has a major role to play in many physical and psychiatric disorders, and it is crucial that clinicians are aware of their potential". In the first part of the sentence the author clearly refers to a placebo effect and in the second the use of placebos. Raz et al. (2008), on the other hand, write that "most practitioners loosely regard the placebo effect as any treatment that improves a symptom or disease but lacks specific effectiveness for the condition being treated". Here 'placebo effect' is used in a narrow sense to refer to something that is usually called 'a placebo'.

Another common source of misunderstanding is the use of phrases like 'use' or 'utilize' or 'employ' the placebo effect. Huculak (2014), for example, writes about "using placebos (or the placebo effect)" [brackets original]. The clinician's aim is to help patients, but at the same time she wonders, "is the placebo effect worth my time and effort". Huculak (2014) also notes that there are "no formal protocols or guidelines available in Canada or the UK for physicians interested in deliberately harnessing these effects in their practices". Klinger and Colloca (2014) even see a "paradigm shift toward the explicit use of the placebo effect".

Cooperman (1999) wonders what a physician should do when "a situation arises that seems to call for the use of the placebo effect". Instead of referring the patient to a practitioner of (so-called) alternative medicine, perhaps physicians should "use the placebo effect" themselves? Cooperman goes on to suggest that "it is acceptable for us to use the placebo effect provided there is no more than a negligible risk of harm and the cost of the 'therapy' is insignificant".

A third and related mistake is to assume, deliberately or not, that the use of placebos is a necessary condition for a placebo effect. According to Foddy (2009), "the placebo effect has been used in clinical medicine, however infrequently, for centuries". However, "the question of whether or not such treatments can form an effective treatment is a source of enduring controversy". Ernst (2008) writes that "even though we now condemn the use of placebo outside clinical trials, most clinicians employ the placebo effect in ways which are not always apparent". Writing about paediatric migraine, Faria et al. (2014) suggests that "recent progress in placebo research is starting to provide clinicians with the tools to introduce placebos and maximize placebo effects in the clinic in an ethically acceptable way, and both doctors and patients seem ready to embrace it". The authors seem to think that placebos must be used to obtain a placebo effect.

Negative language is another cause of misleading interpretations of placebo effects (Bishop et al. 2012). Placebos are described with terms like 'inert', 'dummy' or 'sham', and placebo effect as 'suggestion', 'noise' or 'bias' (Louhiala et al. 2015). Expressions such as "it is merely a placebo effect" are common when denoting a positive outcome of a treatment which the speaker does not consider theoretically plausible. The negative connotations of 'placebo' are transferred into 'placebo effect', although the meaning of the latter is, at least in the clinical context positive.

Problematic distinctions

Miller and Brody (2011) published in 2011 an important paper "Understanding and Harnessing Placebo Effects: Clearing Away the Underbrush", where they "investigate critically seven common conceptual distinctions that impede clear understanding of the placebo effect: (1) verum/placebo, (2) active/inactive, (3) signal/noise, (4) specific/nonspecific, (5) objective/subjective, (6) disease/illness, and (7) intervention/context". They argue that "some of these should be eliminated entirely, whereas others must be used with caution to avoid bias".

For practical reasons, the authors do not suggest elimination of the term placebo effect itself because it has become entrenched and is "unlikely to be abandoned in the near future". Let us examine each of the seven distinctions.

Verum versus placebo. In clinical trials the medication or method under investigation is often called 'verum' (Latin, meaning 'something that is true'). It is thus a general term for something that is compared to placebo in the trial. While there is nothing wrong with the term as such, its etymology may enhance the negative connotations related to 'placebo' and 'placebo effect'. Whatever the form of the placebo in the trial is, it should have no effect at the *practical* level. Labelling one method 'true' and the other 'untrue' or 'false', is, however, misleading. This is more serious if the negative connotation is transferred to 'placebo effect' in the clinical context.

Active versus inactive. According to the most common definition, placebos are inactive (or inactive *enough*, for practical purposes). As we have seen above, the scientific community often defines the concept more broadly, including parts of the placebo effect in the definition. To avoid confusion, Miller and Brody suggest that we not make the active/inactive distinction.

Signal versus noise. In science in general, a signal refers to the phenomenon under investigation and noise to factors that interfere with detection of the signal. In a typical clinical trial, a placebo effect would thus be a typical example of noise. In the clinical practice, however, placebo effects are anything but noise. The same is true in research focusing specifically on placebo effects. The signal-noise distinction is not meaningful for example in research using the open-hidden paradigm described in Sect. 1.1.

Specific versus nonspecific. The distinction between the specific effect of a drug (or method) and the unspecific effect of a placebo is still commonly used although its problems were demonstrated already in the 1980's. Ernst (2008), on the one hand, argued that "if placebos generate clinical benefit, it is imperative that patients profit from the non-specific and specific effects of treatments". Meissner et al. (2013) on the other hand opined that placebo controls are important in research because they "separate specific and nonspecific effects (including placebo effects, regression to the mean, natural course of the disease".

As these examples show, both terms are vague and ambiguous (Alfano 2015a), or "semantic traps for anyone trying to articulate what we do and do not know about placebo effects" (Brody 2009). Basically, 'specific' translates to 'we think we know the mechanism' and 'nonspecific' to 'it seems to have an effect but we don't know the mechanism'. In that sense 'nonspecific' is "at odds with our knowledge of the neurobiological and psychological causes of some placebo effects, which are quite specific indeed" (Alfano 2015a). I agree with Miller and Brody (2011), according to whom "the specific/nonspecific distinction says much more about our prejudices about what counts as 'real' medical knowledge than it does about the workings of placebos".

Objective versus subjective. The best evidence of placebo effects comes from research on subjective outcomes and pain is the most widely studied phenomenon. However, a growing number of studies have demonstrated measurable changes in objective outcomes like various blood markers and physiological responses. Even with respect to pain, placebo effects are not purely subjective, which has been demonstrated by finding the neurochemical mediators of these effects. Miller and Brody (2011) write that "one hypothesis concerning the scope and power of the placebo effect is that it works to change the illness experience of patients rather than affecting the course of disease".

Disease versus illness. A person may have a disease without knowing it and without having any symptoms. On the other hand, they may feel ill but not have a condition that would be classified as a disease by medicine. The disease/illness distinction resembles closely the objective/subjective distinction and both have their roots in the problematic distinction between body and mind. As long as we do not have a language for the unity of 'body' and 'mind' we cannot help but use the dichotomies.

At this stage of research, it seems obvious that clinical placebo effects can be strong with respect to illness experience. However widely placebo effects are understood, very probably they are not strong and long-lasting with respect to the course of diseases.

Intervention versus context. The least problematic and, in fact, the most fruitful distinction is between the actual intervention and the context of care. The open-hidden experiments demonstrate simply that mere information about the administration of a painkiller has a significant effect on the pain experience. This effect is produced by the "simulation or ritual of treatment and the associated context that produces clinical improvement by some psychological mechanism such as expectation or conditioning" (Miller and Brody 2011).

The case of antidepressant trials is a vivid demonstration of the effect of the physician and the context. It is widely acknowledged that there is little difference between the antidepressant groups and the placebo groups in the final outcome (Moncrieff 2015). The scientific debate concentrates on the meaning of the small difference but here it is interesting to note the magnitude of the effect in the placebo groups. Typically, 5 out of 10 subjects appear to respond to active treatment, while 4 out of 10 appear to respond to placebo (Healy 2008). A part of the response in both groups is probably due to spontaneous remission but another part can be due to the effect of the context and the researcher. According to Jakovljevic (2014), "it's not that antidepressants don't work, it's that everything works, including antidepressants".

2.3 New Terminology—Suggestions and Problems

There have been several attempts to clarify the issue of the placebo effect by renaming it. Below, I will discuss four suggestions that have been elaborated in more detail by the original authors: remembered wellness, meaning response, contextual healing and care effect.

Remembered wellness

One of the first attempts to rename the placebo effect was made by Benson and Friedman in 1996, when they suggested that "current levels of clinical efficacy and efficiency could be increased if the placebo effect were reintegrated into routine medical care and reconceptualised as 'remembered wellness'" (Benson and Friedman 1996). The authors paid attention to the heavily negative connotations of the placebo terminology and emphasized the distinction between a placebo and the placebo effect.

The authors chose the term because "ultimately the evocation of the placebo effect depends on central nervous system events that result in feelings of well-being" (ibid.). According to them, "remembered wellness has always been one of the physician's most potent therapeutic assets" and it has "withstood the test of time and continues to be safe and inexpensive" (ibid.).

Benson and Friedman exaggerate their case when they write about the 'reintegration' of the placebo effect into routine medical care, as if it had been 'disintegrated' from it at some point of time. They also use vague language and are uncritical towards some earlier claims: "The placebo effect yields beneficial clinical results in 60–90% of diseases that include angina pectoris, bronchial asthma, herpes simplex, and duodenal ulcer" (ibid.).

The authors do not deepen their suggestion with further analysis, and their suggestion remains somewhat cryptic. They, however, pave the way to a later, more nuanced critique of the prevailing placebo terminology.

Meaning response

Stewart Wolf made a pioneering experiment on placebo effects in the 1940's and 1950's. He wrote about the "pharmacology of placebos" (Wolf 1950, 1959) but understood also the wider context of placebo effects: "the mechanisms of the human body are capable of reacting not only to direct physical and chemical stimulation but also to symbolic stimuli, words and events which have somehow acquired special meaning for the individual" (Wolf 1950).

A more detailed suggestion for new terminology came 52 years later from the American anthropologist Daniel Moerman, who, together with Wayne Jonas, a professor of Family Medicine at Georgetown University, suggested that much of what is called the placebo effect is a special case of the "meaning response", which is defined as the physiological or psychological effect of meaning in the origins or treatment of illness (Moerman and Jonas 2002). When such effects are positive, they include most of the things that have been called the placebo effect and, when they are negative, they include most of what has been called the nocebo effect. Meaning response is attached to the prescription of active as well as inert medications and treatments.

Moerman and Jonas also describe the confusing terminology and note the expansion of the concept of the placebo effect "very broadly to include just about every conceivable sort of beneficial biological, social, or human interaction that doesn't involve some drug well-known to the pharmacopoeia" (ibid.).

The authors refer to several empirical studies in order to demonstrate that although placebos cannot do anything by themselves, their *meaning* can. Words are powerful and affect the world in powerful ways. Denying that could only come from someone "who has never been told "I love you" or who has never read the reviews of a rejected grant proposal" (ibid.).

Most elements of the practice of medicine are, indeed, meaningful. The physician's costume, style, manner and language carry meanings and diagnosis as such may carry meaning far beyond the biomedical construct in question. Surgery is particularly meaningful, as the classical studies of ligation of the bilateral internal mammary arteries as a treatment for angina pectoris (Cobb et al. 1959) and more recent studies on arthroscopic partial meniscectomy for a degenerative meniscal tear (Sihvonen et al. 2013) demonstrate. Patients receiving sham surgery did as well as those receiving the active procedure in the trials.

Moerman and Jonas also refer to the classical study of the correlation between longevity and strength of commitment to traditional Chinese culture (Phillips et al. 1993, see Sect. 1.1.)

Like Benson and Friedman, the authors claim that "we have impoverished the meaning of our medicine to a degree that it simply doesn't work as well as it might any more" (ibid.). They do not, however, present any evidence for this historical claim. Neither do they provide a technical definition of 'meaning', which makes the concept 'meaning response' "inescapably vague" (Annoni and Blease 2018; Chiffi et al. 2019).

Contextual healing

The importance of context was also noticed by Stewart Wolf in the 1950's. In 1959 he wrote that "placebos produce effects on biological mechanisms independently of their chemical properties" but also that "placebo effects derive from the significance to the patient of the whole situation surrounding the therapeutic effort" (Wolf 1959).

Miller and Kaptchuk were the first to develop the idea of context more systematically. They acknowledged in 2008 the dramatic increase in the scientific interest in placebo effects but wrote also that a "major barrier to clinical translation of the substantial investment in laboratory experimentation on the placebo effect is the confusing and misleading way in which this phenomenon is conceived" (Miller and Kaptchuk 2008). They also remarked that "the art of medicine, as reflected in the therapeutic potential of the clinical encounter, has been marginalized in the wake of tremendous advances in the science and technology of medicine" (ibid.).

Because of the confusing and negative connotations of the current terminology the authors suggested reconceptualization of the placebo effect as "contextual healing". By this they refer to the context of the clinical encounter, as distinct from the specific treatment interventions containing factors such as the environment of the clinical setting, the communication between patient and clinician and the rituals of treatment. The context of the clinical encounter should be seen as a potential enhancer or even the primary vehicle of therapeutic benefit. This could be particularly relevant for chronic conditions like pain.

As the experiments in the *open-hidden paradigm* demonstrate, a placebo is not needed to produce a placebo effect and a remarkable part of the therapeutic benefit associated with medication derives from the ritual of the clinical encounter. Empirical studies and theoretical understanding in contextual healing could and should help to isolate and elucidate the factors in the doctor-patient relationship that are related to better outcomes for patients. Placebos are a methodological tool in this research.

Care effect

Tudor Hart and Dieppe wrote in 1996 a paper titled "Caring Effects", in which they emphasized the importance of caring, which has "been central to medical practice in all cultures throughout history, and still motivates most health workers" (Tudor Hart and Dieppe 1996). They cited many research papers, among them a randomised trial according to which patients with symptomatic osteoarthritis randomly allocated to telephone contact fared better than controls (Weinberger et al. 1989). According to the authors, "doctors, nurses, and administrators have been encouraged to give higher priority to technical procedures and pharmacological interventions than to human relationships" (Tudor Hart and Dieppe 1996). They also pointed out that to "account for clinical improvement solely by placebo effects (as simplistic interpretation of controlled trials might easily do) is an insult to the importance of caring" and that "these are caring effects, and so they should be called, as key words in the titles of scientific papers (ibid.).

Louhiala and Puustinen (2008) developed the idea further and suggested that 'placebo effect' should be replaced with 'care effect' to address the outcome of a therapeutic encounter that cannot be attributed to the specific physiological response to the treatment given. They also suggest that the concept of placebo should be limited to the research context only. The term placebo would refer, thus, only to the procedures that are used as inert controls for so-called active treatments in medical research. When inert or only vaguely effective substances or treatments are used in a clinical context, they should not be called placebos. If a method of treatment is ineffective in its own right, let it be called an ineffective treatment for that particular patient or problem. There is no reason to adopt a particular concept for that in a therapeutic process.

According to Louhiala and Puustinen, a care effect is necessarily present in any therapeutic encounter, providing the patient is not unconscious. Whatever the treatment, the patient is being treated. When a patient reports subjective responses that cannot be fully explained in terms of the supposed mechanisms of the treatment given, that may be considered to be the effect of being treated or cared for. Just being acknowledged, heard, understood, assured and comforted can be very alleviating in itself. This can be considered, ultimately, a substantial part of any treatment, irrespective of the caregiver's frame of reference, be it that of a scientifically educated physician, an alternative therapist or an indigenous healer.

A care effect may be small or even negative with respect to the therapeutic intention, but it can never be excluded from the therapeutic encounter. A care effect may be present in a clinical trial, too, but the clinical context and the research context are fundamentally different settings. If present in a clinical trial, a care effect can be considered a confounding factor, not an ally as in clinical medicine.

The authors argue further that the term placebo has many unscientific and pejorative connotations in the clinical context, which have infiltrated also into the concept of placebo effect, regardless of what has been written about the importance, existence and supposed mechanisms of it. For example, it is common to use expressions such as "it is merely a placebo effect" when reference is made to the positive outcome of a treatment that the speaker does not consider theoretically plausible. In contrast, the terms care and caring have neutral or even positive connotations (e.g., of concern and responsibility). They are free of the burden that placebo and placebo effect have acquired as referring to something that is somehow not real. The term care effect respects the reality of the effect of care and suggests a need for more research into the multitude of phenomena behind it.

Louhiala and Puustinen conclude that, in the clinical context, 'placebo effect' could in most cases be readily replaced by 'care effect' without causing any damage to the basic argumentation. That replacement would, in fact, clarify the discussion, since the concept care effect is free of the connotations and contradictory meanings of the term placebo.

Blease (2012) criticizes 'care effect' for being too broad and including "the 'nocebo effect', no effects and the 'placebo effect'". For her a more appropriate successor for 'placebo effect' would be 'positive care effect'.

Between old and new terminology

Two philosophers of science have taken different positions on retaining the concepts placebo and placebo effect. Robin Nunn (2009) suggests that we should "put the placebo construct out of our misery". Jeremy Howick (2017) points out that "the fact that a concept is ambiguous is not, in itself, a sufficient reason for removing them from our vocabulary".

Both Nunn and Howick comment also on the suggested new terms and find them unsatisfactory. According to Nunn (2009), it's not enough "to tweak the placebo construct a little bit here and there with equally mysterious subconstructs such as characteristic factors or to paint it over with a fresh new label such as 'remembered wellness', 'non-specific effects' or 'meaning effects'". For him, "rebranding is not enough to rescue this tired product. Instead, it's time to return to the fundamentals. One fundamental may be that nobody knows enough yet about what's really happening. Mind and body remain deeply mysterious".

Howick (2017) has equally criticized earlier attempts to replace placebo effect with 'meaning response' and 'context effects'. Neither he nor Nunn however do justice to the attempts to clarify the issues with new terminology. Contrary to what Howick and Nunn imply, the earlier authors have not suggested that 'meaning' or 'context' should replace 'placebo' as a prefix to 'effect'. Instead, the new terms describe the phenomena broadly from a different perspective.

For Howick, the strategy is "to try again: to try to produce an acceptable account of placebos that does not fall prey to linguistic confusions" (Howick 2017). He may be successful in providing a useful definition of a placebo in a *narrow* sense, for the purpose of constructing placebo controls in clinical trials. From there it is, however, a long way to a *general* definition of a placebo. Considering the difference between the clinical trial context and clinical practice context I argue that it is not possible to provide a *general* definition of a placebo that would also make sense as a prefix in 'placebo effect'.

Instead of trying to save the concept, Nunn (2009) suggests that we abandon it altogether. In his new world, "a post-placebo era, experiments will simply compare something with something else. That is, they will compare experimental conditions: one group gets these conditions and another group gets those conditions. The report of every methodologically acceptable experiment will describe the conditions that have been compared, so that anyone reading the report may try to replicate them. There will be no hiding behind the skirts of the emperor's new placebo". A similar suggestion was made already in 2002 by Hróbjartsson who wrote that "it might be time to stop using the term placebo effect and instead specify which kind of intervention one is referring to, and how its effect was measured" (Hróbjartsson

2002). It is obvious that both Nunn and Hróbjartsson refer to the clinical trial context and in that narrow sense their suggestion makes sense.

Later in his paper Nunn is, however, less clear. In the following, he seems to refer to a broader concept that includes also the 'placebo effect': "If we put the placebo construct out of our misery, the implications and opportunities are huge. We need new literature, new textbooks, new training, and new laws that expunge the notion of placebo and replace it with something more fundamental, or we admit that we just don't know. Look clearly at the naked emperor and see the body beneath the nothing that covers it. Why wait?" (Nunn 2009).

The task of philosophers is to look far and be as precise as possible. In this case Nunn looks far and Howick tries to be precise. Both have made proposals that are constructive but not readily applicable in the real world of clinical medicine. Meanwhile, the growing community of researchers on placebo effects acknowledge the conceptual difficulties but continue to use the terms in ways that are imperfect and often confusing, which is not necessarily a major problem as such (Blease 2018). A little step forward has been taken, however: when talking about this phenomenon, some researchers have begun to use the plural, acknowledging the fact that there are many different forms of placebo effects.

References

Alfano, M. 2015a. Placebo effects and informed consent. *American Journal of Bioethics* 15: 3–12.

Alfano, M. 2015b. Response to open peer commentaries on "Placebo Effects and Informed Consent". *American Journal of Bioethics* 15: W1–W3.

AMA (American Medical Association). Use in clinical practice. https://www.ama-assn.org/delivering-care/use-placebo-clinical-practice. Accessed 1 Jun 2018.

Andersen, L.O. 2005. A note on the invention, invisibility and dissolution of the placebo effect. *Gesnerus* 62: 102–110.

Annoni, M., and C. Blease. 2018. A critical (and cautiously optimistic) appraisal of Moerman's "Meaning response". *Perspectives in Biology and Medicine* 61: 379–387.

Aronson, J. 1999. Please, please me. *BMJ* 318: 716.

Barnhart, K.T. 2014. Placebo effect in fertility: Advantageous or false advertisement? *Fertility and Sterility* 101: 36–37.

Beecher, H.K. 1955. The powerful placebo. *JAMA* 159: 1602–1606.

Benedetti, F. 2009. *Placebo effects—Understanding the mechanisms in health and disease*. Oxford: Oxford University Press.

Benson, H., and R. Friedman. 1996. Harnessing the power of the placebo effect and renaming it remembered wellness. *Annual Review of Medicine* 47: 193–199.

Bhugra, D., A. Ventriglio, A. Till, et al. 2015. Colour, culture and placebo response. *International Journal of Social Psychiatry* 61: 615–617.

Bird, Alexander., and Tobin, Emma., 2017. Natural kinds. In *The Stanford encyclopedia of philosophy*, ed. Edward N. Zalta, Spring 2017 Edition. https://plato.stanford.edu/archives/spr2017/entries/natural-kinds/. Accessed 1 Jun 2018.

Bishop, F.L., E.E. Jacobson, J.R. Shaw, et al. 2012. Scientific tools, fake treatments, or triggers for psychological healing: How clinical trial participants conceptualise placebos. *Social Science and Medicine* 74: 767–774.

Blease, C. 2012. The principle of parity: the 'placebo effect' and physician communication. *Journal of Medical Ethics* 38(4): 199–203.

Blease, C.R. 2018. Psychotherapy and placebos: Manifesto for a conceptual clarity. *Frontiers in Psychiatry* 379.

Bootzin, R.R., and E.T. Bailey. 2005. Understanding placebo, nocebo, and iatrogenic treatment effects. *Journal of Clinical Psychology* 61(7): 871–880.

Braillon, A. 2009. Placebo is far from benign: It is disease-mongering. *American Journal of Bioethics* 9: 36–38.

Brody, H. 2000. The placebo response. *Journal of Family Practice* 49: 649–654.

Brody, H. 2009. Medicine's continuing quest for an excuse to avoid relationships with patients. *American Journal of Bioethics* 9: 13–15.

Chan, T.E. 2014. Regulating the placebo effect in clinical medicine. *Medical Law Review* 23: 1–26.

Chaput de Saintonge, D.M., and A. Herxheimer. 1994. Harnessing placebo effects in health care. *Lancet* 344: 995–998.

Chaucer, G. 2001. *The merchant's prologue and tale*. Cambridge: Cambridge University Press.

Chiffi, D., A.V. Pietarinen, A. Grecussi. 2019. Meaning and affect in placebo effect. *Journal of Medicine and Philosophy*, in press.

Cobb, L., G.I. Thomas, D.H. Dillard, K.A. Merendino, and R.A. Bruce. 1959. An evaluation of internal-mammary artery ligation by a double blind technic. *New England Journal of Medicine* 260: 1115–1118.

Cooperman, B. 1999. An, "Old Timer" reflects on the use of the "Placebo Effect". *Western Journal of Medicine* 170: 235–236.

Coxe, J. 1808. *The Philadelphia medical dictionary*. Philadelphia: Thomas Dobson.

Cullen, W. 2018. *Clinical lectures*. Edinburgh Feb-April 1772, 218–219. http://www.jameslindlibrary.org/cullen-w-1772/. Accessed 1 Jun 2018.

Duncan, A. 1770. *Elements of therapeutics*. Edinburgh: Drummond; 1834. Editorial. *Edinburgh Medical and Surgical Journal* 42: 483.

Edwards, M. 2005. Historical keywords: Placebo. *Lancet* 365: 1023.

Ernst, E. 2008. A historical perspective on placebo. *Clinical Medicine* 8: 9–10.

Ernst, E., and K.L. Resch. 1995. Concept of true and perceived placebo effects. *BMJ* 311: 551.

Faria, V., C. Linnman, A. Lebel, and D. Borsook. 2014. Harnessing the placebo effect in pediatric migraine clinic. *Journal of Pediatrics* 165: 659–665.

Foddy, B. 2009. A duty to deceive: Placebos in clinical practice. *American Journal of Bioethics* 9: 4–12.

Gold, H., D.B. Barr, M. Cattell, E.F. DuBois, P.A. Bunn, and W. Modell (eds.). 1947. *Cornell conferences on therapy: Use of placebos in therapy*. New York: Macmillan.

Gold, A., and P. Lichtenberg. 2015. Clinical placebo can be defined positively: Implications for informed consent. *American Journal of Bioethics* 15: 25–27.

Gøtzsche, P. 1994. Is there logic in the placebo? *Lancet* 344: 925–926.

Graves, T.C. 1920. Commentary on a case of hystero-epilepsy with delayed puberty. *Lancet* 196:1134–1135.

Grünbaum, A. 1986. The placebo concept in medicine and psychiatry. *Psychological Medicine* 16 (1): 19–38.

Healy, D. 2008. Ethics and science of placebo-controlled trials. *Journal of Psychopharmacol* 22: 598–599.

Hooper, R. 1798. *A compendious medical dictionary*. London: Myrray and Highley.

Hooper, R. 1811. *A compendious medical dictionary*. London: Longman.

Howick, J. 2017. The relativity of 'placebos': Defending a modified version of Grünbaum's definition. *Synthese* 194: 1363–1396.

Howick, J., F.L. Bishop, C. Heneghan, J. Wolstenholme, S. Stevens, F.D.R. Hobbs, et al. 2013. Placebo use in the United Kingdom: Results from a national survey of primary care practitioners. *PLoS ONE* 8 (3): e58247.

Hróbjartsson, A. 2002. What are the main methodological problems in the estimation of placebo effects? *Journal of Clinical Epidemiology* 55 (5): 430–435.

Hróbjartsson, A., and P.C. Gøtzsche. 2001. Is the placebo powerless? An analysis of clinical trials comparing placebo with no treatment. *New England Journal of Medicine* 344: 1594–1602.

Huculak, S. 2014. The placebo effect in psychiatry: Problem or solution? *Journal of Medical Ethics* 40: 376–380.

Jakovljević, M. 2014. The placebo-nocebo response in patients with depression: Do we need to reconsider our treatment approach and clinical trial designs? *Psychiatria Danubina* 26: 92–95.

Jutte, R. 2013. The early history of placebo. *Complementary Therapies in Medicine* 21: 94–97.

Kelley, J.M. 2018. Lumping and splitting: Toward a taxonomy of placebo and related effects. *International Review of Neurobiology* 139: 29–48.

Kerr, C.E., I. Milne, and T.J. Kaptchuk. 2008. William Cullen and a missing mind-body link in the early history of placebos. *Journal of the Royal Society of Medicine* 101 (2): 89–92.

Kienle, G.S., and H. Kiene. 1997. The powerful placebo effect: Fact or fiction? *Journal of Clinical Epidemiology* 50: 1311–1318.

Klinger, R., and L. Colloca. 2014. Approaches to a complex phenomenon—The basic mechanisms and clinical applications of placebo effects. *Zeitschrift für Psychologie* 222 (3): 121–123.

Louhiala, P. 2009. The ethics of the placebo in clinical practice revisited. *Journal of Medical Ethics* 35: 407–409.

Louhiala, P., and R. Puustinen. 2008. Rethinking the placebo effect. *Medical Humanities* 34: 107–109.

Louhiala, P., H. Hemilä, and R. Puustinen. 2015. Impure placebo is a useless concept. *Theoretical Medicine and Bioethics* 36: 279–289.

Louhiala, P., and R. Puustinen. 2017. Meaning and use of placebo: Philosophical considerations. In *Handbook of the philosophy of medicine*, ed. T. Schramme and S. Edwards. Dordrecht: Springer.

Louhiala, Pekka. 2012. What do we really know about the deliberate use of placebos in clinical practice? *Journal of Medical Ethics* 38: 403–405.

Martial. Epigrammata, Book 3, Poem 51. http://www.perseus.tufts.edu/hopper/text?doc=Perseus:text:2008.01.0506:book=3:poem=51&highlight=placebo%2Cplaceam. Accessed 1 Jun 2018.

Meissner, K., M. Fässler, G. Rücker, J. Kleijnen, A. Hróbjartsson, A. Schneider, et al. 2013. Differential effectiveness of placebo treatments: A systematic review of Migraine Prophylaxis. *JAMA Internal Medicine* 173: 1941–1951.

Miller, F.G., T.J. Kaptchuk. 2008. The power of context: Reconceptualizing the placebo effect. *Journal of the Royal Society of Medicine* 101: 222–225.

Miller, F.G., L. Colloca, and T.J. Kaptchuk. 2009. The placebo effect—Illness and interpersonal healing. *Perspectives of Biology and Medicine* 52: 518–539.

Miller, F.G., and H. Brody. 2011. Understanding and harnessing placebo effects: Clearing away the underbrush. *Journal of Medicine and Philosophy* 36: 69–78.

Moerman, D.E., and W.B. Jonas. 2002. Deconstructing the placebo effect and finding the meaning response. *Annals of Internal Medicine* 136: 471–476.

Moncrieff, J. 2015. Antidepressants: Misnamed and misrepresented. *World Psychiatry* 14: 302–303.

Motherby, G. 1785. A new medical dictionary or general repository of physic. Containing an explanation of the terms and a description of the various particulars relating to anatomy, physiology, physic, surgery, materia medica, pharmacy &c. &c. &c. Printed for J. Johnson. London.

Motherby, G. 1791. A new medical dictionary or general repository of physic. Containing an explanation of the terms and a description of the various particulars relating to anatomy, physiology, physic, surgery, materia medica, pharmacy &c. &c. &c. Printed for J. Johnson, London.

Nunn, R. 2009. It's time to put the placebo out of our misery. *BMJ* 338: b1568.

Oswald, F.L. 1902. Cosmopolitan health studies. *Sanitarian* 49: 500–505.

Petronius. Satyricon. http://sacred-texts.com/cla/petro/satyrlat/satl130.htm. Accessed 1 Jun 2018.

Phillips, D.P., T.E. Ruth, and L.M. Wagner. 1993. Psychology and survival. *Lancet* 342: 1142–1145.

Pigg, W.B. 1900. A plea for the better study of diseases of women by the general practitioner. *International Journal of Surgery* 13: 236–238.

Puustinen, R. 2011. Is it psychosomatic?—An inquiry into the nature and role of medical concepts, Durham theses, Durham University. http://etheses.dur.ac.uk/657/.

Puustinen, R., and P. Louhiala. 2013. The paradox of placebo—Real and sham in medicine. In *The medical humanities companion volume three: Treatment*, ed. P. Louhiala, I. Heath, and J. Saunders. Oxford: Radcliffe.

Raicek, J.E., B.H. Stone, and T.J. Kaptchuk. 2012. Placebos in 19th century medicine: A quantitative analysis of the BMJ. *BMJ* 345: e8326.

Raz, A., E. Raikhel, and R. Anbar. 2008. Placebos in medicine: Knowledge, beliefs, and patterns of use. *McGill Journal of Medicine* 11 (2): 206.

Robertson, W. 1776. *Observationes miscellaneae inaugurales De vino praecipue*. Edinburgh: Balfour et Smellie.

Schedlowski, M., P. Enck, W. Rief, and U. Bingel. 2015. Neuro-bio-behavioral mechanisms of placebo and nocebo responses: Implications for clinical trials and clinical practice. *Pharmacological Reviews* 67: 697–730.

Scott, W. 1824. *St. Ronan's well*. Edinburgh: Archibald Constable.

Seneca. De Consolatione ad Helvium. Book 11, Chapter 4. http://www.perseus.tufts.edu/hopper/text?doc=Perseus:text:2007.01.0017:book=11:chapter=4&highlight=placebo. Accessed 1 JUn 2018.

Shapiro, A.K., and E. Shapiro. 1997. *The powerful placebo—From ancient priest to modern physician*. Baltimore: The Johns Hopkins University Press.

Sherman, R., and J. Hickner. 2008. Academic physicians use placebos in clinical practice and believe in the mind–body connection. *Journal of General Internal Medicine* 23: 7–10.

Sihvonen, R., M. Paavola, A. Malmivaara, A. Itälä, A. Joukainen, H. Nurmi, et al. 2013. Arthroscopic partial meniscectomy versus sham surgery for a degenerative meniscal tear. *New England Journal of Medicine* 369: 2513–2522.

Stewart-Williams, S., and J. Podd. 2004. The Placebo effect: dissolving the expectancy versus conditioning debate. *Psychological Bulletin* 130(2): 324–340.

Stuart, W.M. 1889. A remarkable case of morphine addiction. *BMJ* 1051–1052.

Sutherland, A. 1763. *Attempts to revive ancient medical doctrines*, xxiii–xxiv. London: A Millar.

Tudor Hart, J., and P. Dieppe. 1996. Caring effects. *Lancet* 347: 1606–1608.

Weinberger, M., W.M. Tierney, P. Booher, and P. Katz. 1989. Can the provision of information to patients with osteoarthritis improve functional status? A randomized controlled trial. *Arthritis & Rheumatology* 32: 1577–1583.

Wolf, S. 1950. Effects of suggestions and conditioning on the action of chemical agents in human subjects: The pharmacology of placebos. *Journal of Clinical Investigation* 29: 100–109.

Wolf, S. 1959. The pharmacology of placebos. *Pharmacological Reviews* 11: 689–704.

Chapter 3
Placebo Effects

Let us leave the conceptual problems aside for a while and keep the phrase placebo effect. Even then there is a multitude of issues to be examined: Do placebo effects exist after all? If they do exist, what is their nature in relation to the human mind and body? How and why have they evolved? Are there placebo effects outside medicine? Is there a certain kind of personality that is prone to placebo effects? Can placebo effects be seen in children and non-human animals? I will begin with the question concerning the existence of placebo effects.

3.1 But Is There a Placebo Effect After All?

The powerful placebo

An early landmark study in placebo research was the meta-analysis "The powerful placebo" by Beecher (1955). In the end of the paper he concluded:

> It is evident that placebos have a high degree of therapeutic effectiveness in treating subjective responses, decided improvement, interpreted under the unknowns technique as a real therapeutic effect, being produced in $35.2 \pm 2.2\%$ of cases.

The methodological shortcomings of Beecher's study have been demonstrated many times (e.g. Kienle and Kiene 1997) but the 35% figure is alive and well. It is not uncommon that medical doctors refer to this figure and say that "approximately one third of patients respond to placebos", or that "one third of the total effect is placebo effect".

The Beecher study was an early meta-analysis, performed decades before meta-analyses became popular in medicine. From that point of view its shortcomings are

© Springer Nature Switzerland AG 2020
P. Louhiala, *Placebo Effects: The Meaning of Care in Medicine*,
International Library of Ethics, Law, and the New Medicine 81,
https://doi.org/10.1007/978-3-030-27329-3_3

understandable but it is unfortunate that the main conclusion has continued to puzzle the scientific community and the public for more than half a century.

Kienle and Kiene's critique

The main problems with Beecher's study were the following: (1) Because there were no 'no-intervention' groups in the original studies, the possible placebo effects could not be distinguished from other explanations such as spontaneous improvement, a regression-to-the mean phenomenon and the impact of additional treatments; (2) Beecher misquoted two thirds of the trials of his analysis (Kienle and Kiene 1997). The starting point of Kienle and Kiene's re-analysis was the following:

> …the criteria for acknowledging a placebo *effect* taken for this paper are as follows: (1) A *placebo* had to be given, (2) The event had to be an *effect* of the placebo treatment, i.e., the event would not have happened without placebo administration. (3) The event had to be relevant for the disease or symptom, i.e., it had to be a *therapeutic* event. [italics added]

The authors claim that the reported outcome in the placebo groups in *all* of the studies in Beecher's meta-analysis could be "fully, plausibly and easily explained *without* presuming any therapeutic placebo effect". They conclude that "the Powerful Placebo [Beecher's term] turns out to be a fiction". They note that "many factors and phenomena have been summed up under the terms 'placebo' and 'placebo effect', without being *placebos* or *effects* of placebo administration". They also criticise the term 'non-specific effect' as a contradiction in itself.

Kienle and Kiene are, of course, right about the conceptual confusion around the concepts. If 'placebo effects' refer to the effects of the administration of placebos in clinical studies, they are also right when they conclude that "the extent and frequency of placebo effect as published in most of the literature are gross exaggerations". Placebos *as such* do not cause anything but the *administration of placebos* is an act that conveys meanings and messages. In a double-blind study the meaning and message do not differ between the study arms but, for example, the side-effects of the study drug may explain the differences in results.

The authors do not pay attention to the fundamental difference between a clinical study and clinical practice but mention 'psychosomatic effects' in passing. Probably they took for granted the effects of doctor-patient interaction on at least the subjective experience of the patient.

In sum, Kienle and Kiene re-analysed Beecher's meta-analysis and hundreds of other articles on placebo and did not find any reliable demonstration of the existence of placebo effects. They suggested that a valid method to investigate possible placebo effects would be to compare the placebo groups and the untreated groups of trials. Such a study was, in fact, published only four years later.

Is the placebo powerless?

In 2001 Hóbjartsson and Gøtzsche published in the *NEJM* (*New England Journal of Medicine*) a study that soon received wide publicity. The title of the paper was "Is the placebo powerless—an analysis of clinical trials comparing placebo with no treatment". The starting point of the study was the same as in the analysis of Kienle and Kiene (1997), a serious suspicion concerning the validity of Beecher's classical analysis. Another motivation was the positive attitude towards the clinical use of placebos in some editorials and articles (e.g. Oh 1994; Brown 1998). The authors acknowledged the difficulties in defining a placebo and decided to define placebo pragmatically "as an intervention labelled as such in the report of a clinical trial".

The analysis was methodologically rigorous and sophisticated, and the authors also discussed carefully the possible biases that might have affected the findings. They summarised the main findings of the study as follows:

> As compared with no treatment, placebo had no significant effect on binary outcomes, regardless of whether these outcomes were subjective or objective. For the trials with continuous outcomes, placebo had a beneficial effect, but the effect decreased with increasing sample size, indicating a possible bias related to the effects of small trials. The pooled standardized mean difference was significant for the trials with subjective outcomes but not for those with objective outcomes. In 27 trials involving the treatment of pain, placebo had a beneficial effect, as indicated by a reduction in the intensity of pain of 6.5 mm on a 100-mm visual-analogue scale.

If the small placebo effect on pain was real, it corresponded to approximately one third of the effects of NSAIDS (nonsteroidal anti-inflammatory drugs), the most commonly used pain killers in clinical medicine. Hróbjartsson and Gøtzsche questioned the clinical relevance of such an effect.

The authors emphasize the fact that they did not study the effect of the patient–provider relationship. They point out also that the therapeutic effect of this relationship may be largely independent of any placebo intervention.

The study gained wide publicity both in the popular media and in the scientific community, but commentators often simplified the story. A piece of news in the *Scientific American*, for example, described the findings nearly correctly but the headline conveyed a distorted message: "Study finds placebo effect is fake" (Franzen 2001). In a *Lancet* Commentary, Stoessl and de la Fuente-Fernández (2004) defended the role of placebo controls in clinical trials but wrote in passing that "it is ridiculous to assert that the placebo effect does not exist", referring to the *NEJM* study. This was, of course, not what the original authors had concluded.

As could be expected, the study was also heavily criticised, both immediately and in later re-analyses of the original material. The critics pointed out, for example, the 'apples-and-oranges problem' of meta-analyses: the separate studies that are lumped together in a meta-analysis should be similar enough to make the meta-analysis meaningful (Mann 1990). Feinstein (1995) was very critical about the misuse of meta-analyses and called it "statistical alchemy for the 21st century". According to him,

> meta-analytic mixtures are usually too heterogenous to be described with only two fruits. Other writers, with lower levels of enthusiasm or reverence, talk about rotten fruits or even less savory substances.

One of the authors of the *NEJM* paper, Peter Gøtzsche, defended broad meta-analyses in an editorial in the *British Medical Journal* the year before the *NEJM* paper was published (Gøtzsche 2000). He cited some examples of such broad meta-analyses that, according to him, demonstrated the usefulness of the approach. One of his examples was the famous meta-analysis of homoeopathic treatments published in 1997 (Linde et al. 1997). The results of that analysis were peculiar: the pooled estimate indicated that homoeopathy was effective but there was "insufficient evidence from these studies that homeopathy is clearly efficacious for any single clinical condition." (ibid.). According to Gøtzsche, the broad approach made a lot of sense, since there is "no sound empirical basis for believing that homoeopathy should be effective for some conditions and not for others". He continues that,

> Patients and clinicians alike are better served by a reliable answer that there is no convincing evidence that a therapeutic principle, or a class of treatments, is effective, than by an unreliable answer that a particular example of that class of treatment is effective for a particular disease.

A detailed discussion about the proper role of meta-analyses in the evaluation of medical treatments is beyond the scope of this book but I'd like to make two more remarks on this issue. First, the whole idea of meta-analyses of homoeopathic trials is highly questionable, because homoeopathy lacks a scientific rationale. Such research is, in my opinion, a waste of resources. Second, Gøtzsche's argument for broad meta-analyses is not valid in the case of placebos: it is plausible that (the administration) of placebos is effective in some conditions but not others (Wampold et al. 2005). Papakostas and Daras (2001) write on this topic:

> Generally, the presence of anxiety and pain, the involvement of the autonomic nervous system, and the immunobiochemical processes are believed to respond favorably to placebo, whereas hyperacute illnesses (i.e., heart attack), chronic degenerative diseases, or hereditary diseases are expected to resist.

The conceptual confusion around the concepts 'placebo' and 'placebo effect' is obvious here. It seems to me that the meaning of the phrase 'respond to placebo' is different for the *NEJM* authors and for the critics. The former, the authors, seem to be considering the possible effect of placebo *as such* while the latter refer to the whole process of administration of placebos.

In the same line Miller (2001), for example, wrote in his critique that "the clinician–patient relationship is widely considered to be one of the main factors contributing to placebo effects". This is certainly true if 'placebo effect' is understood *broadly*, referring to the context of care (context effect, care effect, meaning response etc., see Sect. 2). What Hróbjartsson and Gøtzsche focussed on, was, however, the possible effect of placebos only.

Hróbjartsson and Gøtzsche performed a similar analysis in 2004 and 2010 with new trials added to the original material (Hróbjartsson and Gøtzsche 2010). Their conclusion was the following:

> We did not find that placebo interventions have important clinical effects in general. However, in certain settings placebo interventions can influence patient-reported outcomes, especially pain and nausea, though it is difficult to distinguish patient-reported effects of placebo from biased reporting. The effect on pain varied, even among trials with low risk of bias, from negligible to clinically important. Variations in the effect of placebo were partly explained by variations in how trials were conducted and how patients were informed.

Four years after the *NEJM* study Wampold et al. (2005) published a re-analysis of part of the original studies. The studies were selected on the basis that they concerned disorders that were amenable to placebos and their design was adequate to detect these effects. According to the authors, the results were strikingly different:

> When studies used in the Hróbjartsson and Gøtzsche (2001) meta-analysis were disaggregated based on the adequacy of the design and the degree to which the disorder was amenable to psychological factors, evidence for a placebo effect was indeed found.

The critique by Wampold et al. was, however, weak for two reasons. First, in the background section of their paper they constructed a straw man which they then attacked. They characterised modern medicine as follows:

> For modern medicine and the physiochemical theory on which it rests, treatments involve interventions that affect the anatomy or physiology of the patient, and these changes purportedly benefit the patient. Any effects produced by other aspects of the treatment, including the hope, expectation, remoralization, therapeutic relationship, or the receipt of an explanatory system are often called placebo effects, with the connotation that such effects are unimportant.

Some people within medicine may think like this but it is a gross over-simplification to characterise *the whole of* 'modern medicine' like this. Hróbjartsson and Gøtzsche (2001) stated explicitly that they "reviewed the effect of placebos but not the effect of the patient–provider relationship".

Second, the results of the re-analysis were not essentially different from the original analysis. For trials with continuous outcomes, for example, both research groups found a statistically significant pooled effect. The interpretation was, however, different. Hróbjartsson and Gøtzsche found "little evidence in general that placebos have powerful clinical effects" but Wampold et al. concluded that "the placebo effect is robust and approaches the treatment effect". In their response Hróbjartsson and Gøtzsche (2007) pointed out that the pooled effects were, in fact, small and that there was a considerable risk of bias due to sample size. They conclude that "Wampold et al. (2005) put powerful spin on their conclusion".

Treatment effects versus placebo effects

Howick et al. published in 2013 one more critical analysis titled "Are treatments more effective than placebos? A systematic review and meta-analysis". Its objective was certainly unique: "to test for differences between treatment and placebo effects within similar trial populations". The studies analysed were the same as in Hróbjartsson and Gøtzsche's Cochrane Review (2010) but Howick et al. compared the 'active' treatment groups against the placebo groups in a meta-analysis. Their main conclusion was that placebos and treatments often have similar effect sizes. They also concluded that "placebos with comparatively powerful effects can benefit patients either alone or as part of a therapeutic regime".

The first conclusion is plausible but also trivial. Many studies fail to show a significant difference between the 'active' treatment group and the placebo group, and very often the effect size of the active treatment is small. The second conclusion is misleading and ethically questionable.

First, due to the setting of the study, no conclusions can be drawn concerning the role of placebos in clinical practice. The original studies did *not* test the effect of placebos on a variety of symptoms and therefore nothing can be said about their effect in any particular condition. In addition, general statements like "placebos and treatments often have similar effect sizes" do not imply anything regarding the use of placebos in the clinic.

Second, there are no such things as 'powerful placebos'. Placebos as such are (practically) inert and the possible 'placebo effect' associated with the use of placebos always takes place in a context and that context is the crucial causal element leading to the effect. It is plausible that the placebo *as such* has a minor or insignificant role in the process.

Third, the authors ignore the difference between clinical research and clinical practice. The aim of a clinical study is to increase scientific understanding and the aim of clinical practice is to help the patient. This difference means that, in general, a significantly higher 'placebo effect' (or, in fact, care effect/context effect/meaning response, see Chap. 2.) can be expected in clinical practice (Tausk et al. 2013).

The main finding in the *NEJM* and Cochrane meta-analyses had been that placebo effects in clinical trials were either missing or small. The aim of the authors was *not* to evaluate the usefulness of placebo in clinical practice but to explore the validity of the earlier arguments about powerful placebo effects in clinical trials. The place of placebos in clinical practice is an *ethical* question and empirical science can provide data for or against the arguments but not solve the question.

Conclusion

In sum, it seems to me that much of the controversy around the existence and importance of clinical placebo effects can be traced back to different meanings given to the concepts. Hróbjartsson and Gøtzsche never wrote that placebo effects do not exist. Their study question was narrow and their interest was to explore the potential clinical effects of placebos. In particular, they did not review the effects of the patient-provider relationship.

Placebo effects are real but as could be seen above and also in Chap. 2, there would be much less confusion if they were called something else. Placebos *as such* do not have meaningful effects but the *context* in which they are given *necessarily* has effects. Far more important is that placebo effects are largely independent of any placebo interventions.

3.2   Body and Mind and Placebo Effects

> ...the comprehensive study of the placebo effect must fundamentally reconceptualize the body—not as the passive site of medical intervention, but as the penultimate multisensory organ and the locus of lived experience. (Thompson et al. 2009)
>
> What is clear is that the body-mind is highly integrated, and insults in any location can affect the organism as a whole. The concept that the body and mind are somehow distinct—encoded

in language such that there is no scientific word to effectively connote full integration between the two—is clearly an inadequate picture of the human organism. (Thompson et al. 2009)

American psychiatrist Walter Brown describes an episode of severe back pain which developed after he had been cross-country skiing for a day (Brown 1998):

> Even tying my shoes was agony. Despite my suffering, I knew there was no serious underlying disease, so I was certain I would be back to normal in no time.
>
> But the days wore on with no change. ... After a week, I became desperate. I called my cousin Gary, who is a physical therapist. ...
>
> As usual, Gary was upbeat and authoritative. After taking my history and putting me through some maneuvers, he identified the muscles involved. He told me to ice the area, prescribed a set of exercises to stretch the constricted muscles and suggested that I take ibuprofen. When the consultation was over, I still had the back pain, but I had a technique for relieving it and the conviction that it would improve. Although my back was not yet better, I was.

Brown's description of his experience—"Although my back was not yet better, I was"—is a typical way to separate the subjective and the objective, or the mind and the body. It is common for all of us, not only among lay people and medical doctors but also among placebo researchers. Consider the following:

> A drug produces an effect on the body, while a placebo primarily works on the mind and imagination of the patient (Szawarski 2004)
>
> "... the placebo effect is derived from its influence on the patient's psyche ... (Rogev and Pillar 2013)
>
> A real placebo effect is a psychobiological phenomenon occurring in the patient's brain after the administration of an inert substance. (Benedetti et al. 2011)

For Szawarski, a placebo works *primarily* on the mind. He acknowledges, however, the patient as a "psychosomatic whole, where states of mind are closely related with states of the body" and therefore "it is possible that the impact on the mind, emotions and imagination of the patient, may have a positive therapeutic effect (Szawarski 2004). Rogev and Pillar do not explore the concepts deeper but note shortly that "the placebo effect, known for its medical impact since the 18th century, is the effect of a certain medical therapy resulting from the patient's belief in the therapy" (Rogev and Pillar 2013). This is only partly true since the patient's belief is not a necessary condition for a placebo effect to occur.

Benedetti is a neurophysiologist which explains his focus on the brain that is obvious throughout his writings:

> These intricate psychological factors can be approached through biochemistry, anatomy, and physiology, thus eliminating the old dichotomy between biology and psychology. (Benedetti 2013)
>
> Therefore, maybe paradoxically, the placebo effect and the doctor-patient relationship can be approached by using the same biochemical, cellular and physiological tools of the materia medica, which represents an epochal transition from general concepts such as suggestibility and power of mind to a true physiology of the doctor-patient interaction. (Benedetti 2013)

Benedetti notes that "[placebos] are made of words and rituals, symbols, and meanings" but concludes that "all these elements are active in shaping the patient's brain" (Benedetti et al. 2011).

He cites the definition of Brody (2000), according to which a placebo response is "a change in the body (or the body-mind unit) that occurs as a result of the symbolic significance which one attributes to an event or object in the healing environment". The term 'body-mind unit' is mentioned only in passing and neither Brody nor Benedetti elaborate it in more detail.

One of the subheadings of Benedetti's review article (Benedetti 2013) summarises the focus of his thinking: "Today the Placebo Effect, or Response, Is an Excellent Model to Understand How the Brain Works". He gives, however, full credit to other areas of research, too, and mentions, for example, the concept of *embodiment* introduced by philosophers and anthropologists. Embodiment will be examined in more detail below. But first a little more about the body-mind dichotomy.

Both lay people and physicians often talk about something being 'just a placebo' or having 'only a placebo effect'. These pejorative expressions carry a hidden message: that a placebo effect is 'only in the mind' and that the mind is separate from the body "which is a physiological system on which drugs and other medical treatments are supposed to operate" (Campbell 2009).

This separation of mind and body is deeply rooted in our thinking and our language and it is often labelled as *Cartesian dualism*, referring to Rene Descartes, the 17th century philosopher. However, his ideas concerning the mind-body question were not so original as they had been formulated and discussed before and Descartes merely repeated and defended contemporary Catholic interpretations of the issue (Puustinen 2011).

This dualism is also said to be characteristic of 'folk psychology' and it certainly has an unconscious influence on much of the terminology that is used in the placebo field (Campbell 2009). Philosophers of mind have abandoned dualistic folk psychology long ago but it is deeply embedded in our everyday language and thought (ibid.). Philosophers or placebo researchers are not an exception here.

Phenomenological perspective

Our everyday language may be hopelessly dualistic but there is a branch of philosophy, which, instead of a naturalistic approach, focuses on the *lived experience* of people. This branch is phenomenology, which "privileges the first-person experience, thus challenging the medical world's objective, third-person account of disease" (Carel 2008).

From the phenomenological point of view, illness is not seen merely as a biological dysfunction to be corrected by medicine but "as a way of living, experiencing the world and interacting with other people" (ibid.).

The gap between mental experience and its behavioural manifestation can be better understood if we accept that subjective responses are not two processes—one mental and one bodily—but a single process recognised by others as well as oneself as *meaningful*. We are not, from this point of view, mind plus body, but *embodied subjectivities* (Cahana 2007).

Embodiment refers to "one's lived experience of one's body as well as one's experience of life mediated through the body as this is influenced by its physical, psychological, social, political, economic, and cultural environments" (Nichter 2008).

From the first-person perspective,

> When I feel pain relief, people cannot 'see' my pain relief. Even functional magnetic resonance imaging cannot 'see' my pain relief, but I am in pain relief. I know I am without any need for observing my behavior: just to feel pain relief is to know one is in pain relief. There is not any point talking about the correct or incorrect use of the expression of pain relief. (Cahana 2007)

According to Cassell (2004), the question "how does the mind act on the body" is the wrong question,

> because it presumes that there is a thing called mind which is separate from the body, that the body is passive to the mind, and that part of the mind's essential nature is that it causes changes–even symptoms or sickness–in the body.

Cassell's thesis is, that,

> We are of a piece; anything that happens to one part affects the whole; what affects the whole affects every part. All the parts are interdependent and not one functions completely separate from the rest. ... The body participates in everything a person does–every moment of life until death.

In a human being, everything physical or biological is also social, emotional and purposeful (Vogt 2014). In one word, persons are *embodied*, they not only *have* a body but *are* the body. Another key concept is *meaning*:

> Persons do things; they act, think, have emotions, create music, express love, get sick, urinate, see and feel things, and more. Not one part of these things can be separated from another except artificially. Meanings are essential to everything that persons do. Meanings– the interpretation and labeling of experience (persons, objects, events, relationships, and circumstances) and the assignment of meaning to words and utterances–are an essential feature of all thought–thinking, reasoning, meditation, imagination, fantasy, valuing, and generating emotion. (Cassell 2004)

Meanings do not *affect* the body as a separate thing, meanings *involve* the body (Vogt et al. 2014) and "include physical manifestations as an essential and irreducible part of the meaning of things" (Cassell 2004, p. 241). Here we have a direct connection to placebo effects: as we saw in Chap. 2, they can be seen as a special case of 'meaning response', a term suggested by Moerman.

We know more and more about the neurophysiological correlates of placebo effects but their causal explanation is not found in the brain but, like many other phenomena, "at the level of the context-dependent functioning of patients as persons, and their automatic, instinctual, habitual or volitional activities, including perception itself." (Vogt et al. 2014)

From a reductionist perspective it is impossible to understand that meaning as such would have an effect on the outcome of any treatment. But it does and it may even be "registered directly at the site of the body—bypassing conscious awareness.

Thompson et al. (2009) acknowledge Moerman's 'meaning response' but argue that "symbolic, emotional and aesthetic elements cannot be boiled down to expectation, desire or meaning, per se."

They argue also that Moerman uses the term 'meaning' uncritically and too broadly for everything that cannot be accounted for by the natural history of the disease or the effect of the drug or treatment as such:

> Throughout his work on placebo, all incidental and preliminary elements of therapy are attributed to 'meaning'. Only at the very end of his book does he mention that 'meaning' encompasses a number of complex and varied representations and relationships—identifying the metonymic relationship (part for whole), the iconic relationship (based on resemblance) and the symbolic relationship (an arbitrary relationship between two things).

What is needed, they argue, is the concept of *performativity*, which "is the power of language to effect change in the world: language does not simply describe the world but may instead (or also) function as a form of social action" (Cavanaugh 2015).

Thompson et al. (2009) give a familiar example of a performative placebo effect: a mother's kiss on a toddler's skinned knee. As we all know–or remember from our own experience—this simple act has the power to eliminate pain often immediately. The effect on the experience of pain in a young child is certainly not consciously cognitive and it cannot be explained as a meaning response only.

An example of performativity in medicine is the power of diagnoses and prognoses. The diagnostic process as such may be a powerful therapeutic tool that can shape the meaning of a condition for patients by providing them with an explanatory framework (Thompson et al. 2009). This is also important even when the result is that the patient does *not* have the disease she's been afraid of. It is probable that prognostic judgments can activate processes that influence health outcomes beyond simple forecasts of future health.

Expectation, one of the central psychological mechanisms of placebo effects, is an example of performativity:

> expectations and imaginative speculation are understood as fundamentally necessary real-time activities in order to mobilise the future into the present. (Brown 2003)

> A social performative interpretation points out that expectation is itself constitutive and performative (Brown 2003), in that it affects change on a level that can bypass conscious cognition and operate directly on the body and on social structure. (Thompson et al. 2009)

Performativity "reorganizes social relationships or reorients the self in the world" (Thompson 2009). An illustrative example of this process is given by Nordstrom (1998) who participated in a ceremony conducted for a woman who had been kidnapped by soldiers and held at their base for months during a war in Mozambique.

> She was physically sick and emotionally traumatized. The ceremony actually began days before the public gathering. Community members stopped by her place of residence to bring food, medicines, words of encouragement, and friendship. They helped the woman piece together a bit of decent clothing to wear, and collected bathwater for her. They sat patiently and told her stories of other atrocities: a constant reminder to her that she was not alone or somehow responsible for her plight. On the day of the ceremony, food was prepared, musicians were called in, and a dirt compound shaded by pleasant trees and plants was

swept and decorated with lanterns and cloth. The ceremony itself lasted throughout the night, a mosaic of support and healing practices. A high point was the ritual bath the woman received at dusk. Numerous women picked up the patient, and carefully gave her a complete bath, which was said to cleanse her soul as well as her body. The bathing was accompanied with songs and stories about healing, dealing with trauma, reclaiming a new life, and being welcomed back into the community. The patient was then dressed in her new clothing and fed a nutritious meal. Shortly thereafter, the musicians began a new rhythm of music, and all the women gathered about the patient to carry her inside the hut. There they placed her on the floor and gathered around, supporting her emotionally as well as physically. The women tended to her wounds; they stroked her much like one would stroke a frightened child, and they quietly murmured encouragements and reassurances. After a while, the women began to rock the patient, and lifted her up among them. They held her up with their arms talking of rebirth in a healthy place among people who cared for her, far from the traumas of war and the past. They carried her outside where the community welcomed her as part of it. (Nordstrom 1998)

What does this story have to do with placebo effects? In fact, a lot. Placebo effects take place in the medical context and the context involves rituals, meanings and people who do their best for the good of the patient.

Thompson et al. (2009) introduce also the concept of *internal performativity*, "in which the practice of rehearsing and/or imaging a particular state of health may itself have perlocutionary force in the body". The authors refer to studies showing how, for example, athletes and musicians benefit from mental practice as well as physical practice. It is plausible that internal performativity focused on healing may have favourable effects as well.

Thompson et al. (2009) note that the positive effects of embodiment have received little attention. As an example of the positively embodied effects of direct sensorial experience they describe the case of Clive Wearing, a British singer, conductor and musicologist, who suffers from severe amnesia following herpes encephalitis in 1985. His amnesia is both retrograde–he remembers very few things from his past–and anterograde–he cannot create new memories.

Clive and his wife Deborah had got married a year before the disease changed their lives permanently and rendered him unable to maintain his memory for more than a few seconds at a time.

Deborah wrote in her 2005 memoir, "Forever Today":

His ability to perceive what he saw and heard was unimpaired. But he did not seem to be able to retain any impression of anything for more than a blink. Indeed, if he did blink, his eyelids parted to reveal a new scene. The view before the blink was utterly forgotten. Each blink, each glance away and back, brought him an entirely new view. Clive was unable to remember anything for so much as a second. He had amnesia so all-encompassing, it erased everything instantly. … I tried to imagine how it was for him. Something akin to a film with bad continuity, the glass half empty, then full, the cigarette suddenly longer, the actor's hair now tousled, now smooth. But this was real life, a room changing in ways that were physically impossible. It must have looked as though the world were ending, the earth falling apart. (Wearing 2005)

Clive did remember, however, some fundamental facts about his life. He knew that he was married–and in a way his love for Deborah was as fresh as it had been 20 years earlier. And he had his music—he could conduct and play the piano.

Clive's musical skills and his love for his wife were not the result of conscious, cognitive memories but rather, they were deeply embodied and sensorial (Thompson et al. 2009).

> Clive and Deborah were newly married at the time of his encephalitis, and deeply in love for a few years before that. His passionate relationship with her, a relationship that began before his encephalitis, and one that centers in part on their shared love for music, has engraved itself in him—in areas of his brain unaffected by the encephalitis—so deeply that his amnesia, the most severe amnesia ever recorded, cannot eradicate it. (Sacks 2007)

This case is sad and beautiful–and extreme. But it tells us also something about placebo effects: "re-experiencing healthful sensations, sensory experiences or emotions that have been inscribed directly into the body can bypass conscious awareness." (Thompson et al. 2009). And there's a word for this phenomenon, invented by the French philosopher Maurice Merleau-Ponty: *motor intentionality*.

Motor intentionality

Intentionality, in general, refers to a relationship between mental phenomena and their objects. It is the relationship of being *about* something or intending *towards* something (Carel 2008).

Frenkel (2008) argues that "the most popular accounts of placebo effects are generally based on some traditional conceptions of intentionality that obscure the role of the body." He argues further that "although psychologists were correct to seek an intentional account for most placebo effects, the kind of intentionality built into the expectancy account fails to include the body at the center of such effects."

As a better way to understand placebo effects Frenkel introduces the concept of *motor intentionality*, originally described by Maurice Merleau-Ponty (1962). Motor intentionality is "a bodily way of knowing and being" that refers "to the intentional activities that essentially involve a bodily understanding of the world" (Frenkel 2008). In Merleau-Ponty's (1962) own words, "my body has its world, or understands its world, without having to make use of my 'symbolic' or 'objectifying' function."

The body is not only passively waiting for mental commands but actively engaged in interaction with its environment. This interaction is meaningful and intelligent (Carel 2008). Sensorial experiences (not only sight, smell, taste, touch and hearing but also bodily sensations like nausea and aesthetic experiences like rhythm) can trigger embodied memories of past experiences, including positive and negative healing experiences. Our experience "in and of the world is perceived, first and foremost, by and in the body" (Thompson et al. 2009). The experience of being-in-the-world is primarily 'embodied', and only secondarily translated into conscious meaning (ibid.). Thompson et al. (2009) argue that the comprehensive study of placebo effects must reconceptualise the body, "not as the passive site of medical intervention, but as the penultimate multisensory organ and the locus of lived experience".

Frenkel uses the case of *biofeedback* as a demonstration of motor intentionality. Biofeedback is a "mind–body technique in which individuals learn how to modify their physiology for the purpose of improving physical, mental, emotional and spiritual health" (Frank et al. 2010). During a biofeedback session, instruments measure physiological activity such as brainwaves, heart function, breathing, muscle activity, and skin temperature. These instruments rapidly and accurately "feed back" information to the user. The presentation of this information — often in conjunction with changes in thinking, emotions, and behavior — supports desired physiological changes. Over time, these changes can endure without continued use of an instrument. (AAPB 2018)

Biofeedback is an example where individuals have no explicit idea about how they actually do what they do:

The agent is not just entertaining a goal or outcome and waiting for it to happen. Neither is she acting upon some conceptual representation of her brain or muscles. The neural circuits of biofeedback only exist for the disembodied observer, from a perspective outside the lived experience of actually doing it. (Frenkel 2008)

Frenkel's examples of highly skilled and unreflective acts are grasping doorknobs or swinging a tennis racket. What had to be learned in each was a *bodily understanding* of the world and how to respond to certain situations. He wants to generalise the idea to a bodily understanding of socially and biologically determined space and argues that

our everyday absorbed coping is also a highly skilled way of being-in-the-world, where much of our understanding is in terms of the body, a way of physiologically interacting with our world that is inseparable from the representations that informs it. (Frenkel 2008)

Frenkel argues further that placebo effects resemble the family of bodily activities we would usually label as skilful and unreflective. Concepts or conscious deliberation are not needed to initiate the changes observed as a placebo effect.

Interestingly, the motor intentional framework applies equally to the physician (ibid.). It is plausible that the physician's own motor intentionality is related to engaging the patient's motor intentionality.

Frenkel sees an analogy here: as in the biofeedback example, where someone does not utilise or need a concept of the skilful action in order to do it, the empathic physician ultimately cannot tell another how she empathises (ibid.).

Thompson et al. (2009) argue that direct embodied experience may be a vitally important aspect of healing and take precedence over meaning-making in the healing encounter. They think that interpretations of the placebo effect that focus on the conscious mind are over emphasised, and suggest further that "what is initially embodied and sensorial may, over time, become cognitive, as narrative, explanation and meaning become attached to the experience." (ibid.).

I agree with the claim by Thompson et al. (2009) "that the field of placebo research has grown in richness and sophistication to the point where the language of 'placebo' restricts our ability to think about complex healing." It seems to me that phenomenology may offer perspectives that help us understand better the complex phenomena commonly referred to as placebo effects.

So far have not touched the question of the *origin* of these effect, whatever they are called. That is the topic of the next section of Chap. 3.

3.3   Why Do Placebo Effects Exist?

Whatever the answer is, it is speculative, but an evolutionary perspective is useful in understanding the roots of these phenomena. Biological systems are products of evolution and placebo effects also must have their roots in our developmental history.

"Self-healing" has very long roots, if we look at the processes at the cellular level (Thompson et al. 2009). There are, for example, multiple mechanisms to repair damaged DNA. If cancer is the consequence of the malfunction of these mechanisms, the relatively low incidence of cancer is indirect proof of the effectiveness of them. At the level of the organism, wound healing, recovery from broken bones and immune response are examples of self-healing (ibid.).

The germ of placebo effects in humans is perhaps in the disposition of parents to comfort the distress of their children (Miller and Colloca 2010). The same can be seen in other mammals but the development of language and the cultural emergence of the healer role make the phenomenon of placebo effects more important in the human species.

A few researchers have developed theories about the evolutionary significance of placebo effects. Below, I will briefly describe three such theories, *the internal health management system* by Nicholas Humphrey, the *signaling theory of symptoms* by Leander Steinkopf and the *evolutionary roots of the doctor-patient relationship* by Fabrizio Benedetti.

Internal health management system

People in pain are interested in relief but from the evolutionary perspective pain is not the problem but actually a self-generated *defence* against further injury (Humphrey 2002). In a way pain could thus be seen as a part of a solution. Other symptoms have evolved in similar fashion: fever has probably evolved as a defence system against infections, nausea and vomiting against spoiled or poisonous food etc. These symptoms are not, however, free of cost: pain debilitates and makes the person vulnerable to further threats; fever drains energy; vomiting throws away not only toxins but also nourishment etc.

According to Humphrey (2002),

> There are two reasons for thinking that evolutionary theory may in fact have something important to say here. One reason is that the human capacity to respond to placebos must in the past have had a major impact on people's chances of survival and reproduction (as indeed it does today), which means that it must have been subject to strong pressure from natural selection. The other reason is that this capacity apparently involves dedicated pathways linking the brain and the healing systems, which certainly look is if they have been designed to play this very role.

Humphrey's use of the terminology is confusing when he writes about the "capacity to respond to placebos". What he actually refers to, however, is the capacity to *elicit positive outcomes*. This is in line with another term he uses, namely the *capacity for self-cure*. Especially relevant in this context is that the capacity for self-cure is not always expressed spontaneously, but can be triggered by the influence of a third party.

Humphrey is fully aware of the confusing terminology and the fact that *all* treatments are likely to bring about 'placebo effects'. His own suggestion for a new term is 'hope-for-relief effect' but he does not develop this idea further. It is plausible that 'genuine' treatments bring along the bonus of the hope-for-relief effects simply because it is easy for the patients to believe in them (ibid.).

Humphrey's own suggestion is that, from the evolutionary perspective, the capacity for self-cure may not be adaptive in its own right. Instead, it

> is an emergent property of something else that is genuinely adaptive: namely, a specially designed procedure for 'economic resource management' that is, I believe, one of the key features of the 'natural health-care service' which has evolved in ourselves and other animals to help us deal throughout our lives with repeated bouts of sickness, injury, and other threats to our well-being.

Later (Humphrey 2005) he calls this system "some kind of internal health management system", which exists "to ensure that the way the body responds to any particular threat is nearly optimal." Because of evolution, humans have these management strategies built into their constitution (Humphrey 2002).

This health management system needs all relevant information about the situation, for example the nature of the threat and the prospects for remission, spontaneously or with help from others. The system also asks: what is going to happen next? Sometimes it may be good to ignore the message of pain, for example, and concentrate on something more important. Humphrey (2002) gives an illuminating example:

> Pain. You've sprained your ankle. Question: Is this the defence you really need right now, or on this occasion will it actually do more harm than good? Suppose you are chasing a gazelle and the pain has made you stop – then, fair enough, it's going to save your ankle from further damage even if it means your losing the gazelle. But suppose you yourself are being chased by a lion – then if you stop it will likely be the end of you.

Empirical proof for the health management system comes, for example, from a study by Breznitz (1999), in which human subjects were made to suffer short-lasting pain by immersing one hand in ice-cold water. The researcher then manipulated their expectations about how long the immersion might last.

There were two groups: the *exact information* group, in which the subjects

> were asked to perform the DT [dynamometer test] with their dominant hand for 70 s, and after a brief relaxation period to keep their nondominant hand in the ice cold water for 4 min. A digital clock in front of their eyes precisely indicated the passage of time.

In the *no information* condition, subjects

> were asked to perform both tasks until the experimenter told them that the test was over. That happened after 70 s for the DT and 4 min for the CPT [cold pressor test]. There was no clock in the subjects' room, and their watches were removed prior to the beginning of the DT.

Success in the test was defined by being able to carry out the tasks to their completion. The most important finding of the study was the following:

Whereas in the No Information condition 30% of subjects were able to keep their hand in the ice water for the entire 4 min, in the Exact Information group the success rate was twice as high.

The intensity of the pain was not measured in the study because the interests of the researchers were elsewhere. Concerning Humphrey's idea of a health management system, Breznitz's conclusion is, however, illuminating:

it is quite conceivable that people are capable of fine-tuning the distribution of their resources according to the anticipated duration ahead of them.

For Humphrey (2002), the very existence of placebo effects is firm evidence of a health management system within us. In a later article (Humphrey 2005) he adds a further point related to the cultural evolution of humans and medicine:

But today, the medicalisation of sickness has changed the picture. For it means there will often be a novel and even overriding piece of information to take into account. People have learned – their culture has taught them – that nothing is a better predictor of how things will turn out when they are sick (whether the pain will ease, whether the infection will abate, whether they will be nursed back to health) than the presence of doctors, medicines, and so on. Yet human beings remain tied to their evolutionary heritage. And so, today, the very prospect of medical attention – the patient's belief in it – works its magic for the simple reason, stemming from the general rule above, that for most of human history, once a sick person has had cause to think that he will soon be safe and well, he has had just the excuse he needs to bring on his own recovery as fast as possible.

Signaling theory of symptoms

Steinkopf (2015) acknowledges Humphrey's theory but criticizes it for weak empirical support. His own contribution is the Signaling Theory of Symptoms (STS), according to which,

discernible aspects of an immune response, such as pain, swelling, or nausea, not only serve a defensive and healing function but also a signaling function: symptoms signal the need for care and treatment to potential helpers. Once help and treatment are granted, the signaling function is fulfilled and the symptoms diminish.

From the perspective of evolution, this mechanism may have been a significant advantage in early human societies, when sick people were more dependent on extensive social support than they are today. Steinkopf suggests that

Some individuals might have had a symptom structure that mobilized support more effectively, while others had symptoms that were less effective in mobilizing help. The former had an evolutionary advantage over the latter since, as described above, mobilizing help is crucial for survival. This would result in a selection pressure for more effective cues.

Discernible symptoms serve thus two functions, defence and signaling for help. Placebo effects are triggered by two main factors: (1) care and social support and (2) the patient's expectation of amelioration. Empirical studies on placebo effects provide support for the STS by demonstrating the importance of human presence and touch in the care of patients.

One example is a study on the effects of ultrasound to reduce swellings caused by dental treatment (Hashish et al. 1988). The use of therapeutic ultrasound was at that time common but few placebo-controlled studies had been conducted. Hashish et al. recruited 125 dental patients who had their third molars extracted bilaterally. The patients were randomly assigned to five groups: (1) ultrasound with slow circular movements of the applicator, (2) placebo-ultrasound (with the intensity set at zero) with slow circular movements of the applicator, (3) stationary placebo-ultrasound (with the intensity set at zero), (4) self-massage and (5) untreated controls.

The results were interesting: there were beneficial analgesic and anti-inflammatory effects in all ultrasound groups but 'real' ultrasound and placebo ultrasound were equally effective. The maximum effect was obtained in the placebo ultrasound group without circular massaging with the applicator. Self-massage by the patient did not produce any significant effect.

An effect—in this case a placebo effect—was thus found only when the experimenter was present. This is in line with other findings in placebo research, for example in the *open-hidden paradigm* experiments (see Sect. 1.1). According to Steinkopf, this supports the STS:

> From the point of view of STS, the excessive swelling was a signaling symptom that had fulfilled its function once someone helped the sufferer. The function was not fulfilled when the sufferer administered the treatment alone, in which case the excessive swelling prevailed.

Another example cited by Steinkopf is a study by Kessner et al. (2013), in which intranasal administration of oxytocin was shown to increase the analgesic effect of a placebo ointment applied to the forearm. The administration of oxytocin has been shown to increase trust and in this study it "might have increased the believability of the instructions by the study physician" (Kessner et al. 2013). According to STS, signaling symptoms should continue until the person can trust that help is coming. In this case, the administration of oxytocin probably increased trust towards the study doctor.

A third example is the study by Valentini et al. (2014) in which the effect of observing facial expressions with different emotional content on placebo analgesia was tested. It was found that passive observation of facial expressions with negative (grimacing—'pain') and positive (smiling—'happy') valence content interacted with the placebo analgesia, with both expressions leading to an increase of the analgesic effect. The enhancement of the placebo effect was greatest when observing a smiling face. Again, this seems to make sense in the light of STS:

> a sad face indicates that a potential helper empathizes with the sufferer, which can be seen as a precondition of helping. A happy face can be seen as an indicator that help is very likely. (Steinkopf 2015)

Steinkopf's signaling theory of symptoms does not contradict Humphrey's internal health management system but rather gives additional perspectives to it. It helps to understand the essential role of human relationships in medical care. Steinkopf presents an illuminating scenario that describes the development of a hypothetical patient's symptoms from the perspective of STS:

> Due to a false alarm of the immune system, a person develops symptoms of abdominal pain and discomfort. The person's spouse does not take the complaints seriously, and the doctor

rejects the person's worries about the abdominal pain, a second doctor does the same, and so does a third. In fact, the doctors are right in their diagnoses but they do not satisfy the patient's social need. With each doctor the symptoms worsen the patient's social need. With each doctor the symptoms worsen, because the signal strength does not seem to suffice to convince potential helpers. The person becomes accustomed to the persistently unsatisfied social need and, hence, might end up developing chronic symptoms, even though a simple acknowledgment and social support might have soothed symptoms at the outset. In the end, the person is treated by a spiritual healer, who takes every complaint seriously, if only to earn money from the patient. Here, the person's symptoms are finally acknowledged and treated interpersonally. Finally, the signaling function is fulfilled and the symptoms soothed.

The evolution of the doctor-patient relationship

According to Benedetti (2013), the evolutionary roots of placebo effects—and, in fact, the doctor-patient relationship—are in the development of *grooming* from a simple scratch reflex. While the function of the latter is simply to remove the cause of a noxious stimulus, the former does not require peripheral stimulation of the skin and its biological function is the care of the body surface. In addition to scratching, grooming involves "licking, preening, rubbing, nibbling and wallowing" (ibid.) and higher areas of the central nervous system are needed than for the scratch reflex only.

The next evolutionary leap was *allogrooming*, which means taking care of the skin of others. In addition to skin care, its function is the regulation of social relationships and it has neurobiological correlates at the cerebral cortex.

Benedetti suggests that the act of grooming is an early form of altruistic behaviour. An element of reciprocity can be seen there, since there are otherwise no immediate benefits to the groomer. Early social behaviour developed from grooming in our ancestors, and they began to take care of others who needed help. Archaeological findings have provided empirical support for compassion among hominids.

During the cultural evolution of man, a single member of the group of hunter-gatherers began to take the role of a person who takes care of the sick:

Prehistoric shamanism represents the first example of medical care, which is characterized by a good relationship between the sick and the shaman. The sick trusts the shaman and believes in his therapeutic capabilities; thus he refers to him for any psychological, spiritual, or physical discomfort. In this way, the shaman acquired a more and more central role and a higher social status in any social group across different cultures. While shamanistic procedures are mainly based on religious beliefs and the supernatural origin of diseases, several rational treatments emerged over the centuries. For example, a broken arm or leg was covered in river clay or mud and the cast allowed to dry hard in the sun, animal skin was used for bandages, and surgical procedures, such as skull trepanning, were carried out. The transition from shamans to modern doctors is recent and depended on the emergence of modern scientific methodology. (Benedetti 2013)

3.4 Mechanisms of Placebo Effects

Research on placebo effects has revealed a lot about their mechanisms, both at the psychological level and at the neurophysiological level. The main focus of this book is on conceptual issues and their clarification but a short description of the psychological and physiological mechanisms of placebo effects is helpful in understanding the richness and broadness of the phenomena and the progress of scientific placebo research during the past few decades. Many of the most recent research projects are, in fact, multi-disciplinary, with psychologists and neurophysiologists working together.

Psychological mechanisms

Colagiuri et al. (2015) begin their review paper by stating that "the placebo effect is a fascinating and important psychobiological phenomenon whereby treatment cues trigger improvement." They write further that

> cues provide information about the likelihood of future events based on the events experienced following those cues in the past. In the context of the placebo effect, when a patient encounters a treatment (whether active or placebo), the verbal, contextual, and social cues present cause the individual to recollect the sensations experienced in prior situations, which in turn develops into an expectancy for what is likely to be experienced in response to the current treatment.

A typical feature of learning is generalisation: the cues that trigger placebo effects do not need to be identical to those that have previously been experienced (Colagiuri et al. 2015).

Treatment cues may come in many forms and psychological mechanisms of placebo effects have been classified in different ways. Traditionally, expectancy theory and classical conditioning have been the main frames of reference. The mechanisms are not separate worlds but overlap significantly.

Below, I will briefly describe expectancy theory, learning mechanisms, somatic attention and feedback, mindset theory, computational models and a semiotic approach that combines different explanations into one theoretical framework. The psychological mechanisms of placebo effects should been seen as different perspectives rather than separate explanations.

Expectations

Placebo effects are highly influenced by treatment expectations of the patients and the treating physicians (Jensen et al. 2014). Expectancy effects have been reported in the scientific literature for more than 200 years. To my knowledge, the first detailed description was presented by the German philosopher and psychologist Ferdinand Ueberwasser in his *Anweisungen zum regelmäßigen Studium der Empirischen Psychologie* (Ueberwasser 1787).

In a chapter titled *Wirkungen der Einbildungskraft auf den Körper* (Impact of imagination on the body) he writes:

> If one vividly imagines the taste of a dish whose consumption has caused us sickness, then sometimes new occurrences of a similar sickness arise: There are examples where bread crumbs, taken in the shape of pills have, by means of vivid imagination and expectations, yielded the same effects as the medication itself. (translated in Schwarz and Pfister 2016)

He also gave an early description of a placebo effect, long before the introduction of the term:

that vivid imaginations, confident expectations of recovery or relief, and therefore firm trust in the physician, or in the medication alone, even if the medication is without effect by itself, can sometimes lead to real relief, or even recovery, for the invalid. This is also why the cures of charlatans are sometimes of so fortunate a success with the common folks who engage with their imagination without restraint if only it is well excited. (translated in Schwarz and Pfister 2016)

A hundred years later John T G Nichols of Harvard University wrote on the same topic:

The average patient listens with much more interest to the prescription of his physician than to his directions about hygiene. Expecting good results from the drug, he often imagines that he feels them. So great is the power of hope that, even in incurable diseases, a temporary improvement often follows each new prescription. This power of hope . . . is sometimes used by the educated physician, who calls it 'expectant attention.' (Nichols 1893, cited in Shorter 2011)

In general, an expectation refers to 'a belief that something will happen or is likely to happen' (http://www.learnersdictionary.com/definition/expectation). The origin of expectations can be implicit knowledge about statistical regularities or explicit beliefs about, e.g., oneself, other agents, situations or specific events.

Closely related to expectation is *desire*, which is "the experiential dimension of wanting something to happen or wanting to avoid something happening" (Price et al. 2008). Desire and expectation "also interact and underlie common human emotions, such as sadness, anxiety, and relief" (ibid.). The open-hidden studies are perhaps the most illuminating example of the power of expectations in pain relief. Adding an overt suggestion for pain relief can increase placebo analgesia to a magnitude that is clinically significant (Price et al. 2008).

Humphrey's (2002) definition of a placebo includes expectations as a central element:

"a placebo is a treatment which, while not being effective through its direct action on the body, works when and because:

- the patient is aware that the treatment is being given
- the patient has a certain belief in the treatment, based, for example, on prior experience or on the treatment's reputation
- the patient's belief leads her to expect that, following this treatment, she is likely to get better
- the expectation influences her capacity for self-cure, so as to hasten the very result that she expects."

"A placebo … which … works" in the definition is misleading if we think about placebos in the standard meaning as inert substances. Obviously Humphrey refers to 'the whole package', i.e. the doctor-patient relationship and the context of care. It is also obvious that the four points describe elements that are valid for *all* treatments, not only those involving placebos. The effect of non-placebo treatments is, of course, not dependent on the expectations only.

The clinical relevance of interventions that induce expectations of pain relief was examined in a meta-analysis of 27 studies (Peerdeman et al. 2016). The conclusion was that clinical pain can be relieved with expectation interventions. In particular, verbal suggestions for acute procedural pain were found to be effective. The effect on chronic pain was smaller.

Learning mechanisms

Classical conditioning. An early description of a classical conditioning mechanism to produce a placebo effect was given by the famous Russian psychologist Ivan Pavlov already in 1927 in his book Conditioned Reflexes (Pavlov 1927). The actual experiment was performed by Pavlov's colleague Krylov, who gave subcutaneous injections of morphine to dogs for several days regularly. Morphine caused nausea, salivation, vomiting and sleepiness in these dogs. After 5–6 days, only the approach and touch of the experimenter were enough to produce the same symptoms. Pavlov did not mention the concept of placebo but describes the mechanism:

> The connection between the morphine itself and the various signals may in this instance be very remote, and in the most striking cases all the symptoms could be produced by the dogs simply seeing the experimenter. (ibid.)

It is obvious that in some cases placebo effects are explained by this kind of classical conditioning only: the patient (or study participant or laboratory animal etc.) learns in the process to react to the conditioned stimulus in the same way she responds to the unconditioned stimulus (Alfano 2015). The stimulus in question may take many forms. For example, the qualities of the pill (taste, form, colour, or shape) or the details of the contact (white coat, technical equipment, personality factors etc.) may be associated with symptom amelioration (Benedetti 2013).

A clinically important discovery is that prior experience with an ineffective treatment attenuates the placebo effect (Colloca and Benedetti 2006). Another clinically important finding is that placebo effects can be established following *partial* reinforcement (Colagiuri et al. 2015). Partial reinforcement means that the cue is paired with the relevant outcome in some, but not all, trials. In practice, individuals are thus able to use all available cues, both verbal or contextual, when learning what to expect from a treatment.

It is often thought that conditioning in human beings requires conscious activity. The question about nonconscious conditioning remains partly open but some evidence seems to support its existence (Colagiuri et al. 2015). Jensen et al. (2015) performed an fMRI study to explore the neural pathways involved in nonconscious activation of conditioned pain responses. They found that both conscious and nonconscious presentation of the conditioned cues led to significant placebo and nocebo responses. Increased activation was found on the orbitofrontal cortex of the brain during nonconsciously activated placebo analgesia. Pharmacological conditioning of non-conscious processes, such as hormonal responses, is also capable of inducing placebo effects (Colagiuri et al. 2015).

Social learning. Colloca and Benedetti (2009) studied the role of social learning in placebo analgesia in an experimental setting, in which the subjects learned by observing the analgesic experience of others. They found that observing the beneficial effects in the demonstrator induced substantial placebo analgesic responses in the study subjects. Hunter et al. (2014) compared video-based and live social observation-induced placebo analgesia and found that the magnitude was equal.

It is also known that emotions expressed by other people influence our own pain experience (Colloca 2014). Valentini et al. (2014) found that the observation of facial expressions with different emotional content (grimacing—'pain', and smiling—'happy') enhanced significantly the placebo analgesia. The enhancement of placebo analgesia during the observation of facial expressions was not correlated with personality traits (like empathy) of the study subjects.

Therapeutic context. Research on the effects of context demonstrates that the two main mechanisms of placebo effects—expectations and conditioning–work together. As Carlino et al. (2014) write,

> Several sensory and social stimuli, such as the doctor's words, including their meaning and tone, the hospital environment and the medical facilities 'tell' the patient that a treatment is being performed.

On the one hand, the patient's conscious positive expectations may thus create a positive context for the treatment outcome. On the other hand, unconscious conditioning may play a role at the same time, when, for example, the qualities of a treatment lead to conditioned placebo response, even when the original active substance of the treatment is missing.

Contexts may also carry special meanings. In an interesting experiment, Coan et al. (2006) studied the effects of handholding on 16 women who were subjected to the threat of (a minor) electric shock. Both psychological and neurophysiological (fMRI) responses were recorded. During the experiment the study subject held the hand of her husband, an anonymous male experimenter or no hand at all. The results were clear: holding the spouse's hand decreased the subjective feeling of unpleasantness compared with holding a stranger's hand or no hand at all. Corresponding change in neural activity could be seen in many areas of the brain.

The clinical context of a symptom is also highly relevant for the patient's experience. Cancer-related pain, for example, can be perceived as more unpleasant than postoperative pain (Carlino et al. 2014) and the reason is obvious: the former is associated with the threat of progress of the disease and eventual death while the latter is associated with hope of recovery.

Somatic attention and feedback

Another perspective for the psychological mechanisms of placebo (and nocebo) effects has been titled *somatic attention and feedback*. Alfano (2015) explains their origin as follows:

> Daily experience is a continuous deluge of stimuli. We notice only some of them. We actively attend to even fewer. When we attend, we interpret or construe in particular ways. … One way in which placebogenic analgesia might occur, then, is simply by raising the threshold both for noticing pains and for counting a sensation as pain rather than, for instance, heat or pressure.

An extreme but familiar example of somatic attention is a soldier who is wounded in combat but is able to ignore his pain until the critical moment is over. Another familiar but less extreme example is the influence of parents' attention on the pain experienced by children.

The latter was studied experimentally by Walker et al. (2006) who assessed the impact of parent attention and distraction on symptom complaints by children with and without chronic functional abdominal pain. A group of paediatric patients with abdominal pain and a control group of well children, as well as their parents, were recruited for a study in which pain was simulated with a water load symptom provocation test. In the test the children were asked to drink water until they feel "completely full". The children could not see the amount of water they drank but relied on internal sensations to assess perceived fullness. The test simulates typical abdominal pain but the discomfort produced is less intense than the clinical pain experienced by these children (ibid.). The parents of the children were randomly assigned and trained to interact with their children according to one of three conditions: (1) attention, (2) distraction, or (3) no instruction.

The results were striking:

> Compared to a control condition that did not manipulate parent behavior, children's symptom complaints nearly doubled under conditions of parent attention and were reduced by half under conditions of parent distraction. (ibid.)

This was the first study in which parents' behaviour towards their children was manipulated under controlled conditions. The water load symptom provocation test induced a discomfort below the pain threshold but the results strongly suggest that parental behaviour has a major impact on children's symptoms in the clinical context, too.

Mindsets

Mindsets are "lenses or frames of mind that orient an individual to a particular set of associations and expectations" (Crum and Zuckerman 2017). They may be but are not necessarily grounded in facts, and they are

> biased or simplified versions of what is right, natural, or possible (e.g., "girls are not good in math"; "diabetes runs in my family, so losing weight won't matter"; "stress is going to kill me"; "this treatment will work because I am in good hands"). (ibid.)

Family and friends, culture, media, religion etc. shape our mindsets, which are *necessarily* present: we can modify our mindsets but we cannot get rid of them. Research on mindsets is a growing field and part of that research focuses on health issues.

It is obvious that over the ages good physicians have been able to shape the mindsets of their patients for their benefit – without having any idea of the concept. In particular, mindsets about the capacity to change and mindsets about treatment efficacy are relevant for health care practice (ibid.).

Patients' mindsets about their capacity to change have an impact on their health:

> The mindset that a combination of diet and medication will reduce blood pressure, for example, may be less likely to be adopted or less effective if the patient has a pre-existing

mindset that "heart disease runs in the family, and there is little I can do to change. (Crum and Zuckerman 2017)

Mindsets about the effects of treatment are, in fact, expectations about the effects. Words as such are powerful. As the open-hidden-paradigm studies show, changing the patient's mindset by telling her that she will receive a powerful painkiller will make a difference. In an empirical study, the actual benefits of the same amount of physical activity were greater when people were made aware that "this work is good exercise" (Crum and Langer 2007).

In another empirical study, the expectations of migraine patients were manipulated by varying the labelling of their medication (Kam-Hansen et al. 2014). In six subsequent migraine attacks, the participants received either a placebo or Maxalt (rizatriptan) administered under three information conditions ranging from negative to neutral to positive (told placebo, told Maxalt or placebo, told Maxalt). It was found that pain relief
was greatest when given positive information. To the surprise of the researchers, the efficacy of Maxalt mislabelled as placebo was not significantly better than the efficacy of placebo mislabelled as Maxalt.

Health information (like any other information) provides a mindset through which the individual interprets her current experience and prepares for future experiences. This new mindset influences attention, motivation and also physiology in a manner that confirms her expectations (Turnwald et al. 2019).

Computational models

According to some recent research, at least some placebo effects also correspond to computational rules and models (Peiris et al. 2018). Pain processing can be divided into two different neurophysiological systems, the sensory-discriminative system, which processes nociceptive stimuli, and the cognitive-affective system, which processes the psychological and affective features of pain. Büchel et al. (2014) have suggested that these systems work together, which means that the brain is not passively waiting for nociceptive stimuli but is actively making inferences based on prior experience and expectations. A Bayesian computational model within the predictive coding framework may account for differences in the magnitude and precision of expectations that influence the strength of placebo effects for pain.

Building on modern concepts of perception, Wiech (2016) has further suggested that also pain perception can be "conceptualized as an inferential process in which prior information is used to generate expectations about future perception and to interpret sensory input". We are more likely to perceive sensory information in accordance with our expectations rather than with competing interpretations that violate such expectations.

Semiotic view on placebo effects

An interesting approach, based on the semiotic theory of the American philosopher Charles Sanders Peirce, combines various psychological, anthropological and cultural explanations of placebo effects (Miller and Colloca 2010).

Semiotics in general can be characterised as a branch of philosophy that focuses on meaning and meaningful communication. There are several schools and branches of semiotics, concentrating on a wide variety of topics in almost all areas of life.

Originally, Peirce developed his theory of signs to provide a systematic understanding of logic, but his semiotic theory can be applied to all forms of communication and learning (Miller and Colloca 2010).

Peirce defined a *sign* as follows:

> I define a sign as anything which is so determined by something else, called its Object, and so determines an effect upon a person, which effect I call its interpretant, that the latter is thereby mediately determined by the former. (Peirce 1998, cited in Atkin 2013)

Atkin (2013) provides a simplified explanation:

> What we see here is Peirce's basic claim that signs consist of three inter-related parts: a sign, an object, and an interpretant. For the sake of simplicity, we can think of the sign as the signifier, for example, a written word, an utterance, smoke as a sign for fire etc. The object, on the other hand, is best thought of as whatever is signified, for example, the object to which the written or uttered word attaches, or the fire signified by the smoke. The interpretant, the most innovative and distinctive feature of Peirce's account, is best thought of as the understanding that we have of the sign/object relation.

It is easy to see the relevance of this for placebo effects. A placebo (pill, device, injection etc.) is a sign that conveys information to the patient (or research subject) (Miller and Colloca 2010). Usually, the placebo intervention is provided in a context that includes more signs,

> that convey information with the potential for producing therapeutic (and also counter-therapeutic or nocebo) responses. These include the clinician's white coat, diagnostic instruments, the appearance of the doctor's office or hospital room, the words communicated by the physician, the physician's disposition in listening and responding to the patient, gestures, and touch. The patient does not come to the clinical encounter as a blank slate but with a history of experiences and memories evoked by prior responses to signs related to the milieu of therapy, some of which may influence the way in which the patient processes the information from signs emanating from the present clinical encounter. (ibid.)

Peirce's theory is complex and he classified signs in several ways. One of these classifications was "according to how their object functioned in signification" (Atkin 2013). In this classification there were three types of signs:

> (i) *indices*, signs which are dynamically connected with their objects, and with the senses or memory of the individuals for whom they serve as signs; (ii) *symbols*, signs which refer to the object that it denotes by virtue of a conventional rule, which causes the symbol to be interpreted as referring to that object, as in the use of language; and (iii) *icons*, signs that signify their objects by virtue of a likeness between the sign and the object, such as diagrams, pictures and representations. (Colloca and Miller 2011)

These three types of signs illustrate the nature and formation of placebo effects. In the case of conditioning, the conditioned placebo effect is a response to the *index* sign that triggers the beneficial outcome (e.g., pain relief). An example of *symbols* operating here is communication, verbal or non-verbal. If a trusted doctor in a safe hospital environment confirms for her patient that "this will help you", many symbols operate for the benefit of the patient, even if the 'pure' treatment as such is not particularly effective. *Icons* in this context could be, for example, anatomical diagrams that help the patient understand her condition. Another example would be a placebo effect in a person who is *observing* another person demonstrating a placebo effect to a treatment (Colloca and Benedetti 2009).

Physiological mechanisms

Levine et al. published in 1978 the ground-breaking study in which they demonstrated that the opioid antagonist naloxone inhibits the placebo analgesic response. Numerous studies have later confirmed this finding which indicates an involvement of endogenous opioids in the process. The opioid system is not, however, the only mechanism involved. At least the dopamine, cannabinoid, and cholecystokinin (CCK) systems are involved in the enhancement and reduction of placebo analgesia (Colagiuri et al. 2015). The central brain structures underlying placebo effects include the ventromedial prefrontal cortex (vmPFC), insula, amygdala, hypothalamus, and periaqueductal gray (Geuter et al. 2017). Because of the many mechanisms it is thus meaningful to talk about placebo effects in the plural.

Neuroimaging studies have increased our knowledge on the neurophysiological mechanisms of placebo effects. Functional magnetic resonance imaging (fMRI) measures brain activity by detecting changes associated with blood flow. With fMRI it is possible to localize brain areas and neural networks that are activated or deactivated during treatment. Most imaging studies have been related to experimental or clinical pain but also patients with depression or anxiety have been studied.

Another modern imaging method is positron emission tomography (PET) which is based on the assumption that areas of high radioactivity are associated with brain activity. PET makes it possible to do research on neurotransmitter activity during various procedures. Modern neuroimaging studies have also shown an important and interesting association between the psychological and the neurophysiological mechanisms of placebo effects. For example, endogenous opioids have been shown to play a role in both expectations and conditioning. According to Geuter et al. (2017), the vmPFC is "a core element of a network that represents structured relationships among concepts, providing a substrate for expectations and a conception of the situation—the self in context—that is crucial for placebo effects."

The neural mechanisms of nocebo effects have also been partly described. It is known, for example that cholecystokinin mediates the nocebo effect related to pain and that this effect can be inhibited with proglumide, which is a known cholecystokinin antagonist.

The introduction of modern research tools has stimulated the study of the biology of placebo effects and already thousands of research papers have given a more and more detailed picture of the physiological correlates of placebo effects. Below I list only a few interesting examples of those studies.

Parkinson's disease and dopamine. Because patients with Parkinson's disease had earlier shown response to placebos, de la Fuente-Fernández et al. (2001) performed a PET study in which they showed substantial release of dopamine in the striatum (area of the brain) when placebo was given to patients.

Expectation and naloxone. Placebo response in pain could be blocked by naloxone if it was induced by strong expectation cues, whereas if the expectation cues were reduced, it was insensitive to naloxone. A well-known side-effect of opioids is respiratory depression and placebo-activated endogenous opioids have been shown to produce this, too (Amanzio and Benedetti 1999).

Expectation and dopamine. Because oral caffeine is known to induce dopamine release, Kaasinen et al. (2004) studied dopaminergic effects of placebo caffeine in healthy human subjects. Eight coffee drinkers were examined in a PET study after no treatment and after oral placebo. In the placebo phase of the study the subjects were told that they had a 50% chance of receiving caffeine, although all were given placebo only. It was found that the placebo induced a significant dopamine release in the thalamus area of the brain. Kaasinen et al. (2004) concluded that caffeine expectation induces dopaminergic placebo effects, and that caffeine and placebo caffeine may share some dopaminergic mechanisms of action.

Similarity between a drug effect and a placebo effect. Because placebo analgesia was shown to be partly dependent upon endogenous opioid systems, it was logical to think that there are some similarities between opioid and placebo analgesia. First proof for this was obtained in a pioneering PET study by Petrovic et al. (2002) who demonstrated that both opioid and placebo analgesia were associated with increased activity in the rostral anterior cingulate cortex (rACC) of the brain.

Alzheimer's disease and placebo effects. In Alzheimer's disease (AD), the frontal lobes of the brain are typically affected, with neuronal degeneration in many areas of the cortex. It is therefore reasonable to expect a loss of placebo responsiveness in these patients. Benedetti et al. (2006) studied AD patients and found that the placebo component of the analgesic therapy was correlated with both cognitive status and functional connectivity among different brain regions. The more impaired the prefrontal connectivity was, the smaller was the placebo response. This finding has important clinical relevance for AD patients, who probably need larger doses of analgesics to compensate for the lack of the placebo response. The same may be true for other patients suffering from similar lesions of the prefrontal cortex (Benedetti 2014).

Placebo mechanisms and the effectiveness of antidepressant therapy. Pecina et al. (2015) studied the association between placebo-activated neural systems and antidepressant responses in 35 patients with major depression. They found that placebo-induced endogenous opioid release in certain areas of the brain was associated with better antidepressant treatment response and concluded that placebo responsiveness could indicate the likelihood of responsiveness to therapeutic approaches also.

Placebo response modulates visual cortex activity. Viewing of affective pictures provokes increased activation in the visual cortex of the brain. In a study by Schienle et al. (2014) participants were presented with disgusting, fear-eliciting and neutral pictures both with, and without a placebo, which was presented with the suggestion that it can reduce disgust symptoms. As expected, placebos provoked a decrease in experienced disgust. At the same time, reduced activation of the primary visual cortex was found. The interpretation was that in addition to exerting effects via emotion regulation networks, placebo effects correlated with neuronal changes in early visual areas.

Your pain and my pain. The ability to feel empathy for other people in pain is associated with the activation of brain areas that are also engaged during the first-hand experience of pain. In a two-step study Rütgen et al. (2015) first showed that inducing analgesia also reduces pain empathy. Then they documented that blocking placebo analgesia with naltrexone (an opioid antagonist) also blocked placebo analgesia effects on pain empathy. The authors concluded that empathy for pain is probably grounded in a person's own experiences of pain. In a recent study it was found that hand-holding during a pain stimulus increased brain-to-brain coupling in a network that correlates with the magnitude of the analgesia and the observer's empathic accuracy (Goldstein et al. 2018).

Physician's brain during treatment of patients. Patient-physician interaction is an essential element in clinical placebo effects (Kaptchuk et al. 2008) and the expectations of the physician influence placebo effects as well as the effects of drugs (Gracely et al. 1985). Jensen et al. (2014) opened a new line of investigations by studying physicians' brain activations during patient–physician interaction while the patient was experiencing pain. They showed that physicians, while treating patients, activated the right ventrolateral prefrontal cortex (VLPFC), a region that has also been shown to activate during placebo responses. According to the authors, it is probable that the VLPFC does not directly modulate incoming pain signals but "represents expectancy for relief by exerting control over brain circuitries with neurochemical resources to modulate pain" (ibid.).

The main focus of physiological placebo research has been on neurobiology but cognitive mechanisms related to placebo effects can activate other physiological responses as well. Here are some examples of studies demonstrating such responses.

Classical conditioning and blood glucose. Stockhorst et al. (2000) conducted placebo-controlled experiments in which they studied "the influence of Pavlovian conditioning when insulin and glucose are injected in doses that induce a state of biochemical hypoglycemia and hyperglycemia, respectively". They were able to show a conditioned decrease in blood glucose and a trend for a conditioned baseline insulin increase in the placebo group.

Classical conditioning and immune response. The first observation of a possible placebo effect involving an immunological mechanism was made as early as 1886, long before such mechanisms were understood or the concept of placebo effect had been introduced into the medical vocabulary. MacKenzie (1886) showed that some individuals with flower allergy react to artificial flowers containing no pollen. Several studies during recent years have demonstrated that immune functions can be modified in humans through conditioning procedures. Kirchhof et al. (2018) showed that learned immunosuppressive placebo responses increased the efficacy of immuno-suppressive medication in renal transplant patients who were already receiving that medication.

According to Irene Tracey (2010),

> Pain does not exist 'out there' by itself but is generated within the brain, akin to pleasure, warmth and other experiences felt subsequent to environmental stimuli. The same goes for relief and analgesia. These subjective, private, multidimensional experiences are observed and measured by behavioral responses or by individual reporting.

She notes further that

> although nociception [the nervous system's response to painful stimuli] is usually the cause of pain, it is neither necessary nor sufficient and is very often not linearly related to the resulting pain. This is because of the many factors that influence nociceptive processing along the pathway from the nociceptor to the spinal cord and brain, including peripheral and central sensitization, genetics, cognition and emotions.

Recent empirical research has confirmed Tracey's statement by demonstrating that placebo treatments affect pain via neurophysiological mechanisms that are largely independent of effects on bottom-up nociceptive processing (Zunhammer et al. 2018)

Human experience is not naturally divided into 'psychological' and 'physiological', but at least so far we don't have language to describe the unity of the experience. Research on the psychological and physiological mechanisms of placebo effects has not solved the age-old mind-body question. It has, however, increased our understanding of the complex interplay between these mechanisms.

3.5   Is There a Placebo Personality?

A *placebo personality* refers usually to a personality type that would be particularly prone to react positively to placebos. Occasionally, however, placebo personality has referred to the characteristics of a good physician, who is likely to obtain good results by having such a personality (Tavel 2014). In this chapter, I use the term to refer to the qualities of the patient or study subject only.

An early study on placebo personality

A debate about a placebo personality has been going on since the early days of placebo research in the 1950's. Numerous studies have addressed the question whether *placebo responders* or *placebo reactors* can be identified, and if this is possible, what their psychological characteristics are. The initial motivation was to decrease the risk of bias in placebo-controlled trials. It was thought, for example, that "an effective drug may be wrongly discarded because data have been diluted by inclusion within the test group of a large number of placebo reactors" (Lasagna et al. 1954).

One of the first major studies addressing the psychological characteristics of placebo responders and nonresponders was conducted by Lasagna et al. in 1954. The subjects were patients recovering from major operations and they were given either morphine or placebo (saline) in a randomised order. The study was conducted in two phases and in both of them the researchers were able to find 'consistent reactors', 'consistent non-reactors' and the largest group who reacted inconsistently, i.e., on some occasions placebos seemed to help and on other occasions not. It was probably not double-blind since the report does not mention blinding.

Extensive psychological evaluation was then performed for the reactors and non-reactors, and definite and consistent personality differences between the two groups were found. Differences in attitudes, habits, educational background and personality structure were demonstrated between the groups but there were no differences in the proportions of men and women or in intelligence between reactors and non-reactors.

According to the interviews and psychological testing, the reactors tended to have more somatic symptoms (like headache and gastric complaints) during stress. They were also more anxious, emotionally labile and self-centred. The reactors had less formal education and were more frequently active church-goers than the non-reactors. Intuitive impressions of the researchers as to which patients were reactors were, however, more often wrong than right. Only intensive interviews and testing could differentiate them from non-reactors.

Lasagna et al. summarised their results as follows:

> There is a certain psychologic set which predisposes to anticipation of pain relief and thus to a positive placebo response. The presence of the traits making this set is probably not an all-or-none phenomenon but rather a graded one. Other factors (such as severity of pain) also affect the response to inert agents, and the resultant of these factors, psychologic and non-psychologic, known and unknown, determines whether or not a particular dose of placebo produces an effect in a given patient.

The psychologists who examined the patients were unaware of the nature of the placebo responses but the staff who administered the medication or placebo were not blinded. This can have affected the results but, in general, the conclusion makes sense. The 'reactors' and 'non-reactors' were two ends of a continuum and the largest group, in fact, was the 'inconsistent reactors'. It is also worth noting that this study concerned only people with postoperative pain and the results could not be generalised to other kinds of symptoms.

Recent advances

Since the 1950's numerous studies on placebo personality have been conducted and many of them have indicated certain characters of 'reactors' or 'responders' and 'non-reactors' or 'non-responders'. Sixty years and numerous studies later the picture is a little bit clearer, and also multifaceted. For example, openness, optimism, empathy, extraversion, altruism and low hostility have been linked to greater responding to placebos (Darragh et al. 2014; Vachon-Presseau et al. 2018). These personality traits do not, however, operate independently. Kelley et al. (2009) studied patients with irritable bowel syndrome who took part in a clinical trial in which they were treated with placebo acupuncture in either a warm empathic interaction ('augmented'), a neutral interaction ('limited'), or were on a waiting list only. One of the key findings was that personality influenced the placebo response, but only in the 'augmented' group. Darragh et al. (2014) conclude that

A cluster of traits characterised by behavioural drive, extraversion, optimism and novelty or fun seeking appears to be germane to placebo responsiveness, but contextual stimuli may differentially activate responses from this 'type'.

Darragh et al. (2015) presented a conceptual model that helps to explain, at least partly, the conflicting findings of studies searching for a placebo personality. They apply the term 'permeability' to "describe a perviousness to environmental factors such as treatment rituals and suggestion". There are two facets of permeability: (i) inward orientation and (ii) outward orientation. The former refers to "a tendency to have an internal focus, or a responsiveness to suggestion as it relates to internal states", and the latter to "a permeability to external input via an approach behavioural style" (ibid.). Suggestibility and acquiescence, for example, have been linked to inward orientation and extraversion, optimism, straightforwardness and altruism to outward orientation.

The central idea of the transactional model of Darragh et al. (2015) is that the placebo response depends on the *interplay* between the environmental cues and the dispositional characteristics of the individual. The model is preliminary but it might be useful in further research.

Placebo responders' brains

During the past two decades, neuroimaging or genetic studies have often been combined to research projects that evaluate placebo effects. In this way new correlations have been found between physiological functions and the tendency to respond or not to respond to placebos.

In the context of pain, for example, the roles of the dopamine system and the endogenous opioid system have been well established. The release of the endogenous opioids and dopamine during placebo analgesia has been shown to occur in several brain regions, including the ventral striatum, which is part of the reward system (Jaksic et al. 2013). Peciña et al. (2013) showed that certain personality traits were associated with activations in endogenous opioid neurotransmission and explained 25% of the variance in placebo analgesic responses.

Positron emission tomography (PET) studies have shown that the *nucleus accumbens* (NAc) of the brain has a major role in placebo analgesia. NAc is closely connected to several other areas that are related to placebo analgesia and probably involved in the dopamine-mediated reward system (Tracey 2010). Scott et al. (2007) studied the role of NAc in two separate experiments with 30 healthy human subjects. They found a correlation between NAc activity during monetary reward anticipation in one experiment and placebo responses in another experiment. They concluded that

intrinsic differences in the function of neurobiological mechanisms involved in reward anticipation processing, such as the mesolimbic dopaminergic pathway, explain a substantial proportion of the variance in placebo effects.

Recently, Vachon-Presseau et al. (2018) performed a series of MRI and fMRI studies and demonstrated that subcortical limbic volume asymmetry, sensorimotor cortical thickness, and functional coupling of prefrontal regions, anterior cingulate, and periaqueductal gray were predictive of placebo response in chronic back pain patients.

In 2016 Colloca et al. (2016) showed that arginine vasopressin was able to boost placebo effects in women but not in men. Interestingly, the effect was independent of the estradiol and progesterone levels of the women.

Placebome

Because the neurobiological differences between individuals explain part of the differences in placebo effects, it is obvious that people are also genetically different regarding their responsiveness to placebos. The study of genomic effects on placebo response is recent and only a few reports have so far been published. Hall et al. (2015) summarised the results and defined the *placebome* as

> the hypothesized group of genome-related or derived molecules (i.e., genes, proteins, or miRNAs) that affect an individual's response to placebo treatment.

The placebome is a new scientific term in the growing family of fields with the suffix 'omics'. In general, 'omics' refers to the study of a *totality* of something ('genomics' is study of the whole genome, 'proteomics' is large-scale study of proteins and their functions etc.). The final aim in the search for the placebome is the same as in the decades-long search for a placebo personality: to increase the efficacy of placebo-controlled trials and to improve clinical care.

The list of candidate genes that may be part of the placebome is so far short but growing. The most promising evidence comes from the study of the genetic variation in the dopamine pathway (Hall et al. 2015).

To take just one example, the role of Catechol-O-methyltransferase (*COMT*), an important enzyme in the dopamine pathway, was studied in a subset of patients of 104 patients from a randomised controlled trial in irritable bowel syndrome (IBS) (Hall et al. 2012; Kaptchuk et al. 2008). It was found that the number of certain alleles (methionine alleles in *COMT* val158met) was linearly related to placebo response as measured by changes in IBS symptom severity. The authors concluded that polymorphism of this gene is a potential biomarker of placebo response (Hall et al. 2012).

Considering the variety of placebo effect mechanisms, Hall et al. (2015) recognise that

> The potential complexity of this network is rapidly escalated when one considers that different diseases and different placebo pathways may produce different responses in different patients.

They discuss also the clinical and ethical implications of the search for the placebome. If the genetic profiles of placebo responders can be established, this knowledge might be useful in research. The clinical use of such data would, however, be far more problematic. At least the following ethical questions arise: should physicians test their patients at all? How would such testing impact on the physician-patient relationship? What if genetic placebo response propensity would appear incidentally in genetic testing? How would knowing about one's tendency to react to placebos affect the actual placebo response? (Hall et al. 2015).

Conclusion

The early enthusiasm about finding a placebo personality changed gradually into scepticism, when later research could not confirm the findings of previous studies. It became obvious that it would be impossible to define a *general* placebo personality. There are at least three reasons for this. First, as the early study by Lasagna et al. (1954) had suggested, the traits are distributed on a continuum and any cut-off point on it is arbitrary. Second, most of the studies have been conducted in the context of pain and it is questionable if the findings could be generalised to other contexts (Darragh et al. 2014). Third, we know now that there are many different forms of placebo effects, not just one. Therefore, it is not even meaningful to think that a class of 'general placebo responders' would exist.

A more modest goal can still be worth pursuing, and more recently, the first steps in the search for the placebome have been taken. In the long run it could, perhaps, be possible to find personality traits or genes that would, in *certain specific circumstances*, increase the probability of a *specific form* of placebo response. In the future, this might help to improve the methodology of clinical trials and increase our understanding of the nature of placebo effects in clinical practice (Jaksic et al. 2013).

3.6 What Is There Outside Clinical Medicine?

Research on placebo effects has taken place mostly in the medical context, with real patients and with healthy volunteers. Originally the focus of research was on the experience and treatment of pain but later, also other symptoms and conditions have gained the attention of placebo researchers. Outside the clinical context, an example is the role of placebos in improving physical and cognitive performance. Outside medicine, many similarities with the medical placebo effects have been found, for example, in marketing research and behavioural psychology.

Placebo effects on physical performance

There is firm evidence that placebo effects (and nocebo effects) play a role in physical performance (Benedetti 2014). The psychological mechanisms are the same as in the clinical context, expectation and conditioning (Beedie and Foad 2009), but the physiological mechanisms are not so well understood. Pain tolerance and success in endurance sport are, however, related and pain placebo mechanisms play a role there, too.

Different conditions have been studied, ranging from weightlifting and short anaerobic sprints to long distance aerobic cycling. Astonishing results were reported in some early studies. Ariel and Saville (1972), for example, studied six weight lifters and found 3–10% training gains during the placebo period, when the subjects were informed that they were given anabolic steroids. The sample size was small but the

results were statistically significant. Beedie et al. (2007) found *both* placebo and
nocebo effects in their study of 42 sprinters who were given a placebo capsule but
given either positive or negative information about its likely effects on performance.

According to Ross et al. (2015), "orally administered placebos have been typically shown to improve endurance performance by circa 2% in participants who are
at least moderately well-trained". They point out, however, that the studies have been
performed on individual practicing alone and not under conditions that would resemble real-world competitions. It is well known that performance is often improved in
a head-to-head competition setting (ibid.).

Berdi et al. (2011) performed a meta-analysis of fourteen studies and calculated
a mean effect size of the placebo intervention in the following way:

$$\text{Effect size:} \frac{\text{mean at trial—mean at baseline}}{\text{standard deviation at baseline}}$$

The overall medium effect size in this analysis was 0.4 (95% confidence inter-
val 0.24–0.56) and the authors concluded that "placebo treatments have a small to
moderate effect on sports performance". In the medical context an effect size of
0.4 would be small and often clinically insignificant but in the sport context, where
milliseconds matter, it would be remarkable.

For many reasons the authors were, however, uncertain about the practical implications of the result (ibid.). First, the studies included had methodological shortcomings. Comparing the placebo group to its own baseline is not ideal since a learning effect during the treatment process may lead to overestimation of the placebo
effect. Like in medical placebo research the optimal comparison group would be
a no-treatment group. Second, results obtained in experimental conditions cannot
be generalised to actual competitions, where several external factors complicate the
picture. The importance of the competition, for example, influences the performance.
Third, remarkable individual differences in the magnitude of placebo responses are
as obvious here as in clinical medicine.

Placebo effects on cognitive performance

There is also evidence that placebo effects play a role in cognitive performance and
other cognition-related tasks (Benedetti 2014). Green et al. (2001) studied the extent
of expectancy in the ability of glucose to affect cognitive performance. The subjects
completed four sessions during which they were given a 500 ml drink 30 min before
completing a set of cognitive assessments. The drink contained glucose or aspartame
and the subjects were either accurately informed or misinformed as to the content
of the drink. The most interesting finding was that glucose administration improved
performance in one of the tests (Bakan test) but only in sessions where the subjects
were rightly informed that they would receive glucose and not when they were
(wrongly) told that they would receive aspartame.

Oken et al. (2008) assigned 40 healthy seniors randomly to two groups, the first of
which was given placebos for two weeks and told it was an experimental cognitive
enhancer. The second group served as a control group with no pills. Taking pills

produced cognitive improvement, for example the CERAD Word List delayed recall test which is commonly used in clinical geriatrics.

Parker et al. (2011) studied the effect of a sham cognitive enhancing drug on *prospective* memory (the ability to remember to perform an action in the future). They found that people in the sham drug group performed better on the prospective memory task than those who were given nothing. The researchers concluded that prospective memory can be improved not only by manipulating how people understand the task, but also by manipulating how people understand their own cognitive abilities.

Weger and Loughnan (2013) recruited 40 participants for their study on the effect of a bogus subliminal priming method on cognitive performance in a general knowledge test. The participants were randomly assigned to the placebo group or the control group. In the experimental phase the placebo group was told that they would see the correct answers presented subliminally below their individual attention threshold, just prior to their answering the respective question. They were also told that although they could no longer consciously recognize what was written, their unconscious would still be able to pick up the correct answer. The placebo group scored higher in the test than the control group.

Placebo-like effects outside the medical context

Many studies have been carried out in behavioural psychology on phenomena that have similarities with the medical placebo effects. It has been found, for example, that multisensory information can be used to improve the design of food and beverage products, as well as the design of dining experiences (Reinoso Carvalho et al. 2016). Multisensory integration has been widely studied in interactions between audition, vision, and touch, and recently also flavour. It has been documented that the shape of the food, and even the shape of a plate can influence the perception of taste (Spence 2015). It has also been studied whether sensory interventions can add value to the product or service by influencing a customer's willingness to pay (Reinoso Carvalho et al. 2016).

Reinoso Carvalho et al. (2016), for example, studied the influence of background music on the beer-tasting experience. Groups of customers tasted a beer under three different conditions: unlabelled beer, labelled beer, and labelled beer together with a short clip from an existing song. It was found that the beer-tasting experience was more enjoyable with music than when the tasting was conducted in silence.

In a study by Harrar and Spence (2013), yoghurt was perceived as denser and more expensive when tasted from a lighter plastic spoon as compared to artificially weighted spoons. Also other properties of the cutlery affected people's taste perception.

Chan and Maglio (2019) performed experiments on coffee-related cues and psychological performance. They found that coffee cues prompted participants to see temporal distances as shorter and to think in more concrete, precise terms.

Meiselman et al. (2000) served identical prepared meals in different environments and found significant differences in the rating of the food. Their main finding was that the food served in a restaurant was consistently rated higher that the same food served in a cafeteria. The obvious explanation resembles one of the main mechanisms of placebo effects: consumers rate their *expectations* of the food in addition to its actual properties.

A placebo-like effect has also been found as a result in marketing actions. Shiv et al. (2005) performed a series of experiments in which they investigated the possibility that price discounts give rise to placebo effects by activating response expectations. They showed that

> consumers paying a discounted price for a product (e.g., an energy drink thought to increase mental acuity) can end up deriving less actual benefit from consuming this product (e.g., they are able to solve fewer puzzles) compared to consumers who purchase and consume the exact same product but pay its regular price.

Like in the medical case, the placebo stimulus (price) affected not only perceived quality but also actual quality, that is, the actual efficacy of the product. When the participants were asked after the experiment if the price of the drink affected their workout, nobody answered affirmatively. The authors concluded that the process by which expectations give rise to the observed placebo effects occurs non-consciously. The placebo-like effects stem from activation of expectancies about the efficacy of the product.

Like in medical placebo research, brain imaging studies have been combined to the psychological and behavioural evaluation of placebo-effect-like phenomena. Enax et al. (2015) studied the neural and behavioural processes underlying the influence of Fair Trade (FT) labelling on food valuation and choice. Forty participants valuated products in combination with an FT emblem or no emblem and stated their willingness to pay in a bidding task while in an MRI scanner. It was found that FT-labelled chocolates tasted better and the participants were willing to pay higher prices for FT products. Subjective value (willingness to pay) was correlated with activity in the ventromedial prefrontal cortex of the brain.

3.7   Placebo Effect and Hawthorne Effect

The 'Hawthorne effect' is often mentioned in the context of placebo research, either as a potential confounder or as a component of the range of response in the placebo group. The name refers to a study from 1924 to 1933 of factory workers at Western Electric's Hawthorne Plant in Illinois, U.S.A. (Parsons 1974; Levitt and List 2011) According to the standard story,

> regardless of the changes made in working conditions–more breaks, longer breaks or fewer and shorter ones–productivity increased. These changes apparently had nothing to do with the workers' responses. The workers, or so the story goes, produced more because they saw themselves as special, participants in an experiment, and their interrelationships improved. (Kolata 1998)

This explanation of the effect has been generalised to other areas like medicine where it is usually taken for granted. Sometimes the effect is called 'observer effect' or 'trial effect' but the interpretation is the same: participants in a study may benefit merely by the act of participation, whatever the intervention is. The effect as such has been demonstrated in medical research but the original reference to the experiments at Hawthorne has been shown to be misleading.

Hawthorne effect revisited

When the experiments at Hawthorne were initiated in 1924, their objective was to answer a narrow question: does better lighting enhance worker productivity? Later, a series of other studies were performed, addressing, for example, the effect of changes in breaks and payment incentives on productivity. The results of these experiments had a profound influence on research in the social sciences and from these experiments emerged the concept of the Hawthorne effect (Levitt and List 2011).

Many parts of the original data obtained at Hawthorne have been re-analysed later and the common finding has been that the results have not been as clear-cut as the original researchers claimed. (Levitt and List 2011). Parsons (1974), for example, performed a thorough re-analysis of the Relay Assembly Test Room experiments and found several inconsistencies in the original reporting. He characterised the Hawthorne effect in experimental research as "the unwanted effect of the experimental operations themselves" and defined it "as the confounding that occurs if experimenters fail to realize how the consequences of subjects' performance affect what subjects do" (ibid.).

The first and most influential studies at Hawthorne were the Illumination Experiments, that gave rise to the popular version of the Hawthorne effect, according to which whenever light was manipulated, output increased. The data from these experiments were never formally analysed and were thought to have been destroyed until recently, when Levitt and List (2011) located a microfilm relating to the experiments. They performed a statistical analysis of these data and found little evidence in favour of a Hawthorne effect as commonly described.

According to the authors, the raw data did, however, produce an impression of such an effect on naïve reading. A closer look at the data revealed another and more obvious explanation: the changes in productivity could be accounted for by the fact that the lighting changes were made on Sundays. From other research it is known that productivity is higher on Mondays than on Fridays and Saturdays. According to the original data at Hawthorne, productivity on Mondays was equally high whether or not a lighting change occurred the day before. The researchers obviously demonstrated the *day of week effect* and not the effect that was later named after Hawthorne, the site of the study.

Levitt and List (2011) concluded that

> an honest appraisal of this experiment reveals that the experimental design was not strong, the manner in which the studies were carried out was lacking, and the results were mixed at best.

They did, however, find some weak evidence consistent with subtler manifestations of (so-called) Hawthorne effects in the data. The workers appeared to respond positively to experimentation over a longer time horizon, and tended to respond more acutely to the experimental manipulations than to naturally occurring fluctuations in light.

It is evident that the lesson from the original experiments was not the Hawthorne effect as it is usually described. There are other lessons, however. Perhaps the most important of these was the *power of a good story* (Levitt and List 2011).

Trial effect in medicine

One of the problems related to the concept of the Hawthorne effect is its many meanings. In the medical literature its existence and meaning are often taken for granted without any elaboration. Sometimes researchers use the term Hawthorne effect but refer to something else, like the nocebo effect. An example of this is a study by Cocco (2009) where he defined the Hawthorne effect as the "impact of prejudice".

More often, however, the Hawthorne effect is understood in the medical research context as an 'observer effect' or 'trial effect'. This means that physiological, psychological or behavioural parameters measured in a study can change by merely from the fact that a person is enrolled in a study (Benedetti 2014).

The occurrence of a trial effect has been demonstrated in a few medical trials. Cizza et al. (2014), for example, studied the effects of study participation per se at the beginning of a sleep extension trial between screening, randomisation, and the run-in visit. They found marked improvements in both sleep duration and sleep quality during a period of 2.7 months between screening and randomisation. Self-reported sleep duration increased by an average of 30 min, although no life-style modifications had been recommended at that point. Particularly interesting in this study was that also biochemical changes like decrease in fasting glucose and fasting insulin occurred during this period. The authors conclude that a 'run-in' period, which is standard practice in medical trials, may, in fact, "markedly alter baseline measurements and therefore have important effects on the predetermined sample size and study outcomes".

McCarney et al. (2007) studied the trial effect in a placebo-controlled trial of *Ginkgo biloba* (a commonly used herb to boost cognition) for treating mild-moderate dementia. In the main study *Ginkgo biloba* was not superior to a placebo, but, in addition to randomisation to treatment, participants were randomised to intensive or minimal follow-up. In the intensive follow-up group, patients were assessed comprehensively at baseline and two, four and six months after randomisation, and in the minimal follow-up group shortly at baseline and fully at six months. It was found that more intensive follow-up resulted in a better outcome than minimal follow-up, as measured by the patients' cognitive functioning.

The overall evidence for a general and clinically important trial effect is, however, missing. Braunholtz et al. (2001) analysed 11 studies that had made direct comparisons of trial and non-trial control groups and found that six of the studies reported a statistically significant trial effect. According to the authors, randomised controlled trials "are more likely to be beneficial than harmful", but, "the effect of trials could be very different in other disease areas—especially where the disease processes, treatments, and outcomes are very different."

3.8 Placebo Effects in Children

For many reasons, including ethical ones, much less clinical research is performed on children than on adults. Placebo effects are, however, well known in paediatrics and there is evidence from clinical studies that they are present in a variety of paediatric conditions, including attention deficit hyperactivity disorder (ADHD), asthma, atopic dermatitis, seasonal allergic conjunctivitis, autism, depression, epilepsy, functional gastrointestinal conditions, chronic fatigue syndrome, hypertension, migraine, syncope and acute nonspecific cough (Simmons et al. 2014; Paul et al. 2014).

In general, there is no difference in the drug responses between children and adults but placebo response rates in clinical trials are often higher in children and adolescents (Weimer et al. 2013). However, only a few experimental studies have investigated placebo effects and their mechanisms per se in children or directly compared the effect sizes between children and adults (ibid.). Indirect comparisons are possible and demonstrate that in many clinical conditions children and adolescents are, in fact, more likely to respond to placebos than adults. It is possible that the inverse relationship between placebo response and age may continue into adulthood (Buck 2012). Below I will first present three examples of studies performed on children and then describe the mechanisms and implications of placebo effects in that population.

Examples of studies in children

Sandler and Bodfish (2008) performed a trial that illustrates not only the placebo effect but also the possible effects of open-label placebos in the paediatric population. The study was an open-label prospective crossover trial conducted on 26 children with ADHD. Their age was between 7–15 years and their condition was stable on stimulant therapy. The study subjects were randomly assigned to one of two conditions: "(1) baseline (100%) dose (1 week), then 50% dose (1 week), then 50% dose + placebo (1 week), or (2) baseline (100%), then 50% dose + placebo, then 50% dose." (ibid.). Both the parents and the children were openly told about the inert nature of placebos but the teachers did not know in which group each child was. Commonly-used ADHD rating scales were used in the follow-up and it turned out that ADHD behaviour tended to remain the same when the dose of stimulant medication was reduced with a placebo but to deteriorate when the dose was reduced without a placebo.

Paul et al. (2014) found a significant placebo effect in the treatment of young children with nonspecific acute cough. They recruited 120 children aged 2–47 months and randomised them into three groups: (1) agave nectar, (2) placebo or (3) no treatment. Both agave nectar and placebo proved to be superior to no treatment. The authors concluded that the "parents receive some added level of reassurance by administering a potential remedy to their child" (ibid.).

Case studies are usually considered to give weak evidence of the general effectiveness of therapies but sometimes they can, however, powerfully demonstrate the effect of a treatment on a single individual. Bachiocco and Mondardini (2010) describe the case of an infant affected by a complex cardiac defect and tracheal agenesis. Because of the latter, frequent bronchoscopies had to be performed at 3–6 day intervals. Midazolam and fentanyl were used for sedation and pain relief, but at the age of 7 months the baby began to anticipate the procedure by looking at the personnel's movements and positioning her body ready for the bronchoscopy.

At one of the procedures she assumed the optimal position by herself, then stayed calm, remaining motionless and quiet. The anaesthetist decided not to administer either midazolam or fentanyl and the baby remained calm during the whole procedure. The authors interpreted this as an obvious example of a placebo effect, the mechanism of which was classical conditioning to the repeated procedures. During later bronchoscopies, however, the baby showed signs of distress and was given medication to alleviate her symptoms.

Mechanisms of the difference in placebo effects between children and adults

The age-related special features of children account for the differences in placebo effects between children and adults. Several explanations have been proposed.

First, children have pronounced learning skills and because learning is a major component of placebo effects, this could explain part of the difference (Colloca 2015; Weimer et al. 2013).

Second, the degree of suggestibility appears to be modulated by age (Simmons et al. 2014). According to Parellada et al. (2012), almost 80% of children are considered suggestible compared to 15% in the adult population.

Third, children "seem to have the ability to approach a situation from an unbiased or naive perspective because of lack of experience and perhaps the absence of prefrontal control" (Faria et al. 2014). It is possible that a mature prefrontal cortex hinders flexible thinking, because of prior experience that affects expectations (ibid.).

Fourth, in the case of children, there is not only the dyadic doctor-patient relationship but the triad doctor-parent-child (Colloca 2015; Simmons et al. 2014; Tates and Meeuwesen 2001). The complex relationships within a triad are a potential source for placebo effects in children.

Doctor-parent-child relationships and placebo effects

In a commentary on a trial on amitriptyline in children with functional gastrointestinal disorders, Benninga and Mayer (2009) write:

> in contrast with the straightforward physician patient interactions in the adult, the treating physician exerts a dual effect on the pediatric patient: One directly on the patient, the other on the parent– child 'dyad', thereby indirectly influencing the child's expectation through changing parents' attitudes and expectations.

This indirect effect has been referred to as a *placebo by proxy effect* and defined as a response to therapy that is affected by the behaviour of other people who know that the patient is undergoing therapy (Whalley and Hyland 2013). It is difficult to distinguish the effect of the patient's beliefs from the effect of others' beliefs, which makes it difficult to study the effect in older children (ibid.). This effect may be particularly important in young children and infants, where it cannot be explained by their knowledge of treatment (Simmons et al. 2014). The mechanisms that underlie the usual placebo effects are likely to shape placebo by proxy effects also (Grelotti and Kaptchuk 2011). A few studies in this age group have demonstrated the placebo by proxy effect.

Hoover and Millich (1994) recruited to their study 35 young boys reported by their mothers to be behaviourally 'sugar sensitive'. The boys and their mothers were randomly assigned to two groups. The boys in both groups were given aspartame (an artificial sweetener) but the mothers in the two groups were told different stories. The mothers of the experimental group were told their sons had received a large dose of sugar, but the mothers in the control group were told their sons received placebo. The main result of the study was that mothers in the 'sugar expectancy condition' rated their children as significantly more hyperactive than did control mothers (ibid.). The mothers of the former group also "acted on their expectancies by maintaining more physical proximity to their sons, giving more criticisms, and talking more frequently to them, in an apparently 'hovering' or controlling manner" (ibid.).

Whalley and Hyland (2013) studied 58 children with frequent tantrums reported by their parents. The children were given flower essence, which, according to the man-ufacturer, should reduce temper tantrums but which, according to the researchers, was actually a placebo. The parents were told that the researchers were "investi-gating the effectiveness of a commercially available flower essence", and that the "researchers neither endorsed nor rejected the treatment as a way of helping parents of children with temper tantrums" (ibid.). The parents were, however, given "infor-mation which was identified as coming from those who believed in the usefulness of flower essences" (ibid.). Tantrum frequency, tantrum severity, and parental mood were measured for eight days before treatment and for ten days after the start of treatment. A highly significant continuing reduction in tantrum frequency and sever-ity was found over the days of flower essence treatment. There were also significant day-to-day correlations between parents' mood and tantrum frequency and severity.

It is obvious that placebo by proxy effects are not necessarily objective changes in the child's behaviour. Instead, they can be due to parental perception only. This means that the parents may perceive a change independent of the pharmacologic effects of medication. The parents may also change their own behaviour, which then leads to objective changes in the child's behaviour. Further, the changes in parental perception may alter parental behaviour, which then alters the child's behaviour (Simmons et al. 2014).

Sometimes it may be misleading or even dangerous to draw conclusions on the behaviour of the child. The following example from neonatology shows this.

It is commonly believed that oral sucrose relieves procedural pain in neonates. Several clinical trials seem to demonstrate this but the problem is that the usual outcome measures used in this population are indirect and based on behavioural and physiological observation. Slater et al. (2010) performed a study in which they measured pain-specific brain activity evoked by a heel lance and recorded with elec-troencephalography (EEG). They showed that oral sucrose did not significantly affect activity in neonatal brain or spinal cord nociceptive circuits. The fact that sucrose reduces observed clinical pain scores in newborn infants does not necessarily imply pain relief (ibid.).

Implications of placebo effects by proxy and the higher rate of placebo responses in children

Placebo effects by proxy are not limited to children. Adult patients also very often have, for example, family members or friends, whose interests and experience influence the patient's symptoms. In the best case placebo by proxy effects and placebo effects interact to create positive change (Grelotti and Kaptchuk 2011). If the parents or other family members feel optimistic, their behaviour as such may mediate the placebo effects related to the treatment.

Placebo by proxy may, however, cause harm, in which case it could, of course, be called *nocebo by proxy*. The above example concerning sucrose for neonatal pain serves as an example: the behaviour of the baby after a dose of sucrose may mislead the parents and personnel to think that she has less pain, although this may not be the case. An unjustified course of antibiotics prescribed only because the parents demand it may be harmful for the child even though the parent would report a positive response.

High placebo response rates in paediatric research are also a methodological challenge for medical research. A placebo response in adult migraine trials is around 35% but 50% or higher in paediatric trials (Faria et al. 2014). If the response in the drug group of the trial is not substantially larger than in the placebo group, very large sample sizes are needed to demonstrate statistically significant differences between the groups.

It is usually suggested that the high placebo responses should somehow be eliminated. Faria et al. (2014) take a look in the other direction and make a suggestion:

> …instead of focusing on eliminating placebo responses in clinical migraine trials, the focus should be redirected towards understanding the underlying mechanism responsible for high placebo response rates in children with migraine (after excluding confounding factors such as spontaneous remission) in order to maximize and use them therapeutically.

3.9 Placebo Effects in Non-human Animals

It should not be a surprise that placebo effects are found also in non-human animals. My guess is that veterinarians and animal owners have understood this phenomenon for ages, without, however, referring to the concepts of placebo or placebo effect.

The phenomenon is known in the world of veterinary science also, but not much research exists in this area. The first scientific study in veterinary science that used the term placebo effect was published in 1962 (Hernnstein 1962).

Hernnstein's starting point was the hypothesis that classical conditioning, as demonstrated by Pavlov and others in the early 20th century, may be the most important factor explaining placebo effects in humans. Hernnstein showed in his own experiment that saline injections alone could induce physiological changes in rats, if saline injections were first coupled with scopolamine, an anticholinergic drug. Hernnstein concluded that this

> placebo effect is based on the animal's experience and can be eliminated by withholding the drug, in conformity with the traditional paradigm of simple Pavlovian conditioning. There appears to be no reason to suppose that the placebo effect in human patients differs in any way from that demonstrated here, other than in degree of complexity. (ibid.)

Since placebo by proxy effects are common in children, it is plausible that they exist also in non-human animals. An analogous but non-medical version of placebo by proxy effect in animals was demonstrated for the first time shortly after Hernnstein's original paper. Rosenthal and Fode's (1963) starting point was the *experimenter bias*: "Experimenters can all too easily find what they are looking for (support for their own hypotheses) by inadvertently influencing the way in which their subjects behave" (ibid.).

The researchers recruited students to their study and falsely led them to believe that they were acting as experimenters, while they were actually used as a means to study the experimenter effect on the behaviour of rats. Each student was given five rats which were to be trained on a T-labyrinth task. Part of the students were led to believe that their rats had been bred 'bright' to complete the task in the labyrinth, and part of the students were told that their rat had been bred 'dull', correspondingly. In reality, the rats were standard laboratory rats. As expected, the rats which were believed to be bright made more correct responses and were quicker than the rats which were believed to be dull. The authors concluded that they had demonstrated the experimenter effect in rats and that the effects were probably much more powerful in studies of human behaviour (ibid.).

During the 2000's a few studies have been carried out that have directly addressed placebo effects in animals. In 2012 Conzemius et al. published a study on 'caregiver placebo effect' which they defined as the effect seen after a sham medical intervention "causes pet caregivers (owners or veterinarians) to believe the treatment they provided to the pet improved the pet's condition" Conzemius 2012). The study included dogs in the placebo arm of a larger double-blind placebo-controlled trial of deracoxib in the treatment of lameness due to osteoarthritis in dogs. A large caregiver effect was found for both pet owners and veterinarians: 57% of the owners and 40–45% of the veterinarians reported improvement although no change was found in an objective test for the limb function.

Similar results were found in a meta-analysis of five placebo-controlled studies on degenerative joint disease-associated pain in cats. Responses reported by the owners of the cats were remarkably higher than the change in the measured activity of the cats (Gruen et al. 2017). Interestingly, sometimes there seemed to be a beneficial effect on the cat's activity from the owner's belief that the cat was on active treatment (ibid.).

Munana et al. (2010) performed also a meta-analysis of three placebo-controlled canine epilepsy trials. They found that 79% of the dogs in the placebo groups demonstrated a decrease in seizure frequency and 29% had a 50% or greater reduction in seizures. As in any clinical placebo-controlled study, there are many possible explanations for the change, including spontaneous recovery, regression to the mean and higher level of care related to the study. It is, however, probable that a placebo-by-proxy effect explained part of the change. It was also found that

> when owners believed that their dog was receiving the active treatment, they reported a significantly greater improvement in pain signs compared with those who believed their dog was receiving placebo or were uncertain about the treatment administered. (ibid.)

Conditioning and expectancy are two different mechanisms of placebo effects but there is significant overlap between these models. Non-human animals have the cognitive capacity to form and act on expectations although they cannot express their expectations like humans can.

Perhaps the best example of expectancy effects in non-human animals are *winner and loser effects*. Briefly, "social experience influences the outcome of conflicts such that winners are more likely to win again and losers will more likely lose again, even against different opponents" (Rutte et al. 2006). The ultimate causes of the winner and loser effects are unknown but they are prevalent throughout the animal kingdom. There is evidence for an adaptive advantage for both winners and losers if they behave according to their social experience (Schwartz et al. 2006). According to a study with rats, those who acted upon their prior fighting experience benefitted by reducing fighting costs in subsequent encounters. Rats with a winning experience benefitted by being able to decide the following contest faster and rats with a losing experience benefitted by causing their opponents to be less aggressive, thus reducing their own risk of being injured (Lehner et al. 2011).

The results of all these studies bear a lot of resemblance to the findings of placebo research in humans. They demonstrate various aspects and mechanisms of placebo effects, the placebo-by-proxy effect and the discrepancy between subjective reporting and objective findings. After all, the similarities between human and non-human placebo effects is not a surprise, when bearing in mind their evolutionary roots.

3.10   Nocebo Effect—The Evil Brother

Nocebo effects are usually understood as the opposite of placebo effects, i.e. negative effects that are not causally related to a treatment as such but arise in the treatment or research context. According to Ruan and Kaye (2016),

> Two variants of these nocebo responses exist: one is characterized by new symptoms or a symptom aggravation associated with drug or placebo intake, although the chemical agent itself is not able to trigger these symptoms. Another variation of nocebo responses is the reduced efficacy of clinical interventions due to negative expectations or prior experiences.

The term 'nocebo' is Latin and translates "I shall harm". To my knowledge, it was used for the first time in the medical context in Walter Kennedy's article "The

Nocebo Reaction" (Kennedy 1961). He first described the 'placebo reaction' and 'placebo reactors', patients "whose symptoms are improved by the administration of the 'blank' preparation" (ibid.). Like many others in those days, Kennedy believed that placebo effects are linked with a certain type of personality, the suggestible patient.

Next, Kennedy pointed out that hardly any attention had been paid to the contrary effect, which he—as the first author in the world—called the nocebo reaction. His message to the medical world was that:

> The point of importance is to recognise that nocebo reactions do occur, and not to confuse them with true pharmacological effects, such as the ringing in the ears caused by quinine. (ibid.)

Kennedy gave credit to his predecessors who had described the multitude of 'adverse reactions' in the placebo groups of clinical trials. He also pointed out that "every doctor has met the nocebo reactor, even if he has not labelled him as such" (ibid.). Analogously with beliefs concerning placebo reaction, Kennedy believed that nocebo reaction "refers to a quality inherent in the patient rather than in the remedy" (ibid.).

We have seen earlier that conceptual problems related to the concepts placebo and placebo effect are complex. The conceptual issues related to nocebo and nocebo effect may be equally or even more complex. Look at the definition of nocebo in the Merriam-Webster Dictionary:

> a harmless substance or treatment that when taken by or administered to a patient is associated with harmful side effects or worsening of symptoms due to negative expectations or the psychological condition of the patient (https://www.merriam-webster.com/dictionary/nocebo)

The plausibility of the definition depends on the interpretation of the verb 'associate'. We may assume that it is the context that causes the adverse effects, not the substance or treatment *as such*. As so often in the case of placebo, the concept nocebo is used ambiguously, often referring to the nocebo effect. Although it is a common claim, neither placebos nor nocebos 'cause' or 'produce' anything, but given in a context their use may be related to desirable or undesirable effects.

Nocebo effects are often defined as the opposite of placebo effects. Some authors do, however, provide separate definitions for a nocebo effect or nocebo response:

> a nocebo effect is the induction of a symptom perceived as negative by sham treatment and/or by the suggestion of negative expectations. A nocebo response is a negative symptom induced by the patient's own negative expectations and/or by negative suggestions from clinical staff in the absence of any treatment. (Häuser et al. 2012)

> In contrast to the placebo effect, a nocebo effect or nocebo response encompasses negative rather than positive therapeutic outcomes (adverse events) that follow the administration of an inert treatment. For the neuroscientist or psychologist, the nocebo effect is related to negative expectations of clinical worsening. (Benedetti et al. 2016)

> The nocebo effect is the causation of sickness (or death) by expectations of sickness (or death) and by associated emotional states. Two forms of the nocebo effect should be recognized: In the specific form, the subject expects a particular negative outcome and that outcome consequently occurs; for example, a surgical patient expects to die on the operating table

and does die—not from the surgery itself, but from the expectation and associated affect. In
the generic form, subjects have vague negative expectations—for example, they are diffusely
pessimistic—and their expectations are realized in terms of symptoms, sickness, or death—
none of which was specifically expected. (Hahn 1997)

These definitions resemble the definitions of placebo effects by describing the phenomenon and being vague at the same time. But as we saw earlier in Chap. 2, it
is possible to discuss these issues meaningfully even without precise definitions.
Placebo effect or nocebo effect are not the best names to describe these phenomena
but so far there is no consensus about better, more precise names.

   Nocebo phenomena were first described within an anthropological context, without the word 'nocebo', however (Benedetti 2013). So-called voodoo-death is an
extreme example of the consequences of negative expectations.

   The general conceptual frame for understanding placebo and nocebo effects is the
same, although there are differences in the physiological mechanisms (Colagiuri et al.
2015). There is much less research addressing nocebo effects directly than there is
research on placebo effects, but something is known about the physiology of nocebo
effects, particularly in the area of pain. It has been found, for example, that proglumide, a cholecystokinin antagonist, mediates the nocebo response in which pain
is increased (Benedetti 2013). In contrast with the association of placebo responses
with increased dopamine and opioid activity, nocebo responses are associated with
decreased dopamine and opioid release (Ruan and Kaye 2016).

Empirical studies on nocebo effects

One of the earliest and most widely cited studies on nocebo effects was performed in
Japan by Ikemi and Nakagawa in 1962. They studied 57 high school boys, who were
first asked to complete a questionnaire concerning allergy in general and hypersen-
sitivity to lacquer tree and wax tree in particular. The group was then divided into
four categories according to their answers concerning reactions to lacquer tree or
wax tree ('strong reaction', 'moderate reaction', 'no sensitivity' or 'no experience').
Various experiments were carried out, in which the boys were exposed to either a
poisonous substance from lacquer or wax trees or to a control substance which was
obtained from the chestnut tree. In the experiments the boys' arms were touched with
leaves or raw extracts from the trees. The subgroups of the study were small but the
main finding was consistent: the stronger the reported allergy, the commoner was the
nocebo effect ('effect of suggestion' in the report). Interestingly, skin pathology in
the nocebo effect group was very similar to that induced by contact with the leaves
or extracts from the lacquer and wax trees (Ikemi and Nakagawa 1962).

   One year later Pogge (1963) provided the first comprehensive list of nocebo effects
found in clinical trials. He did not use the word nocebo but titled his paper "The toxic
placebo". In a population of 3549 patients from 67 studies he found 38 different side-
effects attributed to placebo medication, nine of these occurring in more than 1% of
the patients.

   After these early reports nocebo effects have gained more attention only recently.
There are many kinds of nocebo effects as there are different placebo effects, and
therefore it is not meaningful to try to find a *general* estimate of the magnitude of

nocebo or placebo effects. However, in some narrow areas it may make sense to try to quantify the magnitude of placebo or nocebo effects, although it should be kept in mind that combining different kinds of studies is problematic in many ways. The magnitude of adverse effects in placebo-treated patients has been investigated in many patient groups, including Alzheimer's disease, Parkinson's disease, multiple sclerosis, neuropathic pain, fibromyalgia, and migraine (Amanzio et al. 2016).

Several meta-analyses on the magnitude of placebo effects on pain have been published but research on the magnitude of nocebo effects has been less common. An example is a meta-analysis by Petersen et al. (2014), who combined the results of ten individual studies. The overall magnitude of the nocebo effect was moderate to large, as measured with standard statistical tools used in meta-analyses. The findings were similar to those found in meta-analyses of placebo effects in pain (ibid.). The authors conclude that "roughly speaking, it may be equally easy or difficult to obtain nocebo and placebo effects." (ibid.).

In another meta-analysis, Mitsikostas et al. (2014) combined data from 21 randomised placebo-controlled anti-depressant trials. They found that almost half (44.7%) of the placebo-treated patients reported at least one adverse event and that 4.5% discontinued placebo treatment due to adverse effects.

The impact of drug information during administration of an analgesic cream was addressed by Aslaksen et al. (2015) Altogether 142 healthy subjects were randomised into six groups: (1) Emla (a topical analgesic cream) with information about analgesia; (2) Emla with neutral information ("a medical cream"); (3) Placebo with information about analgesia; (4) Placebo with information about hyperalgesia: (5) Emla with information about hyperalgesia and (6) No medication. Experimental pain was induced with a 30*30 mm aluminium contact thermode applied to the right volar forearm, the location of the cream. Pain, subjective stress and blood pressure were measured before, during and after the exposure to the stimulus. The most interesting result was that the pain-relieving effect of Emla could be reversed to its opposite by negative information: while all participants in group 1 experienced decreased pain, 95% of the participants in group 5 (Emla and negative information) experienced increased pain.

The next example is related to the side-effects of statins that are extremely widely used medicines in the prevention of major vascular events like myocardial infarctions and strokes. Their role is particularly clear in secondary prevention but evidence is accumulating for their advantages also in primary prevention. In placebo-controlled randomised trials of statins, side-effects have been rare but according to observational studies, up to 20% of statin users report adverse effects like muscle pain. Gupta et al. (2017) approached this dilemma by analysing the data of a large placebo-controlled trial of atorvastatin in the prevention of heart disease. The main finding of the trial was that cholesterol lowering with atorvastatin reduced the risk of heart disease by about 35%. The study itself was interrupted early at the recommendation of the data safety and monitoring board because of substantial cardiovascular benefits in those assigned atorvastatin. Trial physicians were invited to offer atorvastatin to all study patients until the end of the planned follow-up nearly three years later. The main finding of Gupta et al. was that no excess of muscle-related side-effects was reported in the atorvastatin group during the blinded randomised period, but a significantly higher rate of them were reported in the second, non-blinded, non-randomised part of the study when both physicians and patients were aware that statin therapy was being used (ibid.). In an accompanying editorial Pedro-Botet and Rubiés-Prat (2017) state that:

> the strengths of Gupta and colleagues' study lie in the fact that these were the same patients, no run-in period existed to exclude patients intolerant to therapy, and few patients had previously taken any statins.

Both the original authors and editorialists argue that the results clearly illustrate a nocebo effect: there was an excess rate of muscle-related side-effects only when physicians and patients were aware that statin therapy was being used and not when its use was blinded.

Nocebo effect of informed consent

Respect for patient autonomy and not harming the patient ('non-maleficence' or 'primum non nocere') are two key principles in medical ethics. There are situations where the two principles are in tension or conflict and one of these situations is related to the nocebo effect. It has been suggested that a request for informed consent may sometimes be harmful for the patient (Cohen 2014; Fortunato et al. 2017; Faasse et al. 2019).

A typical example of such nocebo effect is disclosure of information that results in unintended harmful outcomes for the patient. Numerous studies have demonstrated that these effects are real and significant in the research context (Fortunato et al. 2017) and there is no reason to believe that the problem is less important in clinical medicine. A daily dilemma for every doctor working in clinical medicine is the extent of disclosure of potential side-effects of a treatment to the patients.

Various solutions to this dilemma have been presented. One possibility would be an *authorised concealment* approach (Miller and Colloca 2011). This means that patients would be informed about the nocebo effect phenomenon and asked if they

> would prefer not to be informed about potential treatment side effects, such as nausea or headaches, that are psychologically bothersome but do not pose serious risks of harm to health? (ibid.)

Another approach has been suggested by Cohen (2014), according to whom

> nondisclosure should be directly related to the degree of sensitivity of the particular side effect or complication to suggestion. For example, not all symptoms of Parkinson's disease are equally sensitive to suggestion: while bradykinesia is sensitive, tremor and rigidity apparently are not. The more the potential side effect is sensitive to suggestion, the more caution should be applied in disclosing it. (ibid.)

Wells and Kaptchuk (2012) argue for a *contextualized informed consent*, which

> involves taking into account the possible side effects, the person being treated, and the disease involved, to tailor the information provided about medication side effects to provide the most transparency with the least potential harm. (ibid.)

Particular candidates for nondisclosure would, according to Wells and Kaptchuk, be nonspecific symptoms, although the boundary between specific and nonspecific symptoms is, of course, not always clear.

Fortunato et al. (2017) propose that

> nondisclosure of side effects for treatments with nocebogenic potential is ethical on the grounds of nonmaleficence, even at the expense of autonomy qua full informed consent at the initiation of treatment. (ibid.)

They also suggest that some kind of follow-up should be arranged to determine the possible experiences of side-effects.

Let us now consider the suggested solutions in the light of a common clinical example. A family doctor has made a diagnosis of urinary tract infection and prescribes a course of trimethoprim (an antibiotic) to the patient. What should the patient be told about potential side-effects, the list of potential side-effects is extensive, ranging from common (such as nausea, vomiting, itching, and dermatitis) to very rare (like agranulocytosis, anaphylaxis, aseptic meningitis and colitis)? Most of the suggested solutions to the information dilemma are not easily applied in this case.

A suggestion concerning authorised concealment (Miller and Colloca 2011) would sound very strange to an average patient. Cohen's (2014) idea of relating nondisclosure to the degree of sensitivity of the side-effect to suggestion would be impractical also, since in the above case it would imply discussion about agranulocytosis but not nausea. Fortunato et al.'s (2017) term 'treatment with nocebogenic potential' is vague here and a follow-up for the side-effects would also be impractical.

Wells and Kaptchuk's (2012) idea of a contextualized informed consent makes sense in this case, since it allows the information to be tailored individually, taking into account the personality of the patient, the disease and the treatment in question. It also allows for cultural understanding of the patient and her circumstances. The patient's level of health literacy, for example, may be an important factor determining the nature of information provided (Widdershoven et al. 2017).

Clinical relevance of nocebo effects

The clinical significance of nocebo effects is not minor. The occurrence of nonspecific side-effects like nausea, fatigue, dizziness or pain are correlated with physical characteristics of the therapy, dose and duration, but research on nocebo effects suggests that patient expectations play a major role, too (Garg 2011).

There is a lot of research on nocebo effects resulting from information concerning possible side-effects of treatments, and Colloca (2017) suggests the following strategies to minimise harm: (1) avoid negative phrases when explaining interventions or medical treatment; (2) emphasise the positive effects of the medical treatment; (3) take time to allow the patient to ask about previous and actual negative aspects of the medication and (4) provide adequate time for the patient to internalise the information that has been given.

The strategies are certainly relevant for clinicians, but Greville-Harris and Dieppe (2015) go one step deeper and argue that the key factor in understanding nocebo effects is *invalidation*, which means "communicating a lack of understanding and acceptance to the patient". This may, of course, happen unintentionally.

Validation and invalidation are concepts developed for psychotherapy and health care communication in general. Validation refers to "communicating acceptance and understanding to another person" (ibid.) and invalidation is the opposite. Validation is not the same as empathy or compassion but focuses on communicating understanding and acceptance.

The effects of invalidation can be harmful and produce nocebo effects:

> The validation/invalidation research highlights that professionals who believe that they are being compassionate and empathetic with the patient may be invalidating them, because the kindly, reassuring interaction may be interpreted as patronizing or indicative of a lack of belief in the severity of the condition by the patient. For example, reassuring patients that there is nothing physically wrong with them, when they are in a great deal of pain, can leave them feeling misunderstood and delegitimized. (ibid.).

Another perspective on the importance of nocebo effects comes from psychological research comparing the impact of good and bad factors on the lives of individuals. In a thorough analysis of empirical evidence on the topic, Baumeister et al. (2001) conclude that "bad is stronger than good", that is,

> Bad emotions, bad parents, and bad feedback have more impact than good ones, and bad information is processed more thoroughly than good. … Bad impressions and bad stereo-types are quicker to form and more resistant to disconfirmation than good ones. Various explanations such as diagnosticity and salience help explain some findings, but the greater power of bad events is still found when such variables are controlled.

The conclusion may sound pessimistic but Baumeister et al. conclude also that

> This is not to say that bad will always triumph over good, spelling doom and misery for the human race. Rather, good may prevail over bad by superior force of numbers: Many good events can overcome the psychological effects of a single bad one. When equal measures of good and bad are present, however, the psychological effects of bad ones outweigh those of the good ones.

The implications of "bad is stronger than good" in the health care context are obvious. Negative experiences and expectations will probably have greater impact on the patient's well-being than equally strong positive experiences and expectations. Luckily, repeated positive experiences may finally overcome the effects of a single negative experience.

References

AAPB (The Association for Applied Psychophysiology and Biofeedback, Inc.). About Biofeedback. 2018. https://www.aapb.org/i4a/pages/index.cfm?pageid=3463. Accessed 1 Jun 2018.
Alfano, M. 2015. Placebo effects and informed consent. *American Journal of Bioethics* 15: 3–12.
Amanzio, M., and F. Benedetti. 1999. Neuropharmacological dissection of placebo analgesia: Expectation-activated opioid systems versus conditioning-activated specific subsystems. *Journal of Neuroscience* 19: 484–494.

Amanzio, M., S. Palermo, I. Skyt, and L. Vase. 2016. Lessons learned from nocebo effects in clinical trials for pain conditions and neurodegenerative disorders. *Journal of Clinical Psychopharmacology* 36: 475–482.

Ariel, G., and W. Saville. 1972. Anabolic steroids: The physiological effects of placebos. *Medicine and Science in Sports* 4: 124–126.

Aslaksen, P.M., M.L. Zwarg, H.-I.H. Eilertsen, M.M. Gorecka, and E. Bjørkedal. 2015. Opposite effects of the same drug: Reversal of topical analgesia by nocebo information. *Pain* 156: 39–46.

Atkin, Albert, "Peirce's Theory of Signs" The Stanford encyclopedia of philosophy (Summer 2013 Edition), and ed. Edward N. Zalta. https://plato.stanford.edu/archives/sum2013/entries/peirce-semiotics/. Accessed 1 Jun 2018.

Bachiocco, V., and M.C. Mondardini. 2010. Julia's placebo effect. *Pain* 150: 582–585.

Baumeister, R.F., E. Bratslavsky, C. Finkenauer, and K.D. Vohs. 2001. Bad is stronger than good. *Review of General Psychology* 5: 323–370.

Beecher, H.K. 1955. The powerful placebo. *JAMA* 159: 1602–1606.

Beedie, C.J., and A.J. Foad. 2009. The placebo effect in sports performance: A brief review. *Sports Medicine* 39: 313–329.

Beedie, C.J., D.A. Coleman, and A.J. Foad. 2007. Positive and negative placebo effects resulting from the deceptive administration of an ergogenic aid. *International Journal of Sport Nutrition and Exercise Metabolism* 17: 259–269.

Benedetti, F. 2013. Placebo and the new physiology of the doctor-patient relationship. *Physiological Reviews* 93: 1207–1246.

Benedetti, F. 2014. Placebo effects: From the neurobiological paradigm to translational implications. *Neuron* 84: 623–637.

Benedetti, F., E. Carlino, A. Piedimonte. 2016. Increasing uncertainty in CNS clinical trials: The role of placebo, nocebo, and Hawthorne effects. *Lancet Neurology* 15: 736–47.

Benedetti, F., C. Arduino, S. Costa, S. Vighetti, L. Tarenzi, I. Rainero, and G. Asteggiano. 2006. Loss of expectation-related mechanisms in Alzheimer's disease makes analgesic therapies less effective. *Pain* 121: 133–144.

Benedetti, F., E. Carlino, and A. Pollo. 2011. How placebos change the patient's brain. *Neuropsychopharmacology Reviews* 36: 339–354.

Benninga, M.A., and E.A. Mayer. 2009. The power of placebo in pediatric functional gastrointestinal disease. *Gastroenterology* 137: 1207–1210.

Berdi, M., F. Köteles, A. Szabo, and G. Bardos. 2011. Placebo effects in sporst and exercise. *European Journal of Mental Health* 6: 196–212.

Braunholtz, D.A., S.J.L. Edwards, and R.J. Lilford. 2001. Are randomized clinical trials good for us (in the short term)? Evidence for a "trial effect". *Journal of Clinical Epidemiology* 54: 217–224.

Breznitz, S. 1999. The effect of hope on pain tolerance. *Social Research* 66: 629–652.

Brody, H. 2000. The placebo response. *Journal of Family Practice* 49: 649–654.

Brown, W.A. 1998. The placebo effect. *Scientific American* 278 (1): 90–95.

Brown, N. 2003. Hope against hype—Accountability in biopasts. *Presents and Futures. Science Studies* 16: 3–21.

Büchel, C., S. Geuter, C. Sprenger, and F. Eippert. 2014. Placebo analgesia: A predictive coding perspective. *Neuron* 81: 1223–1239.

Buck, M. 2012. The placebo response in pediatric clinical trials. *Pediatric Pharmacology* 18 (3): 1–4.

Cahana, A. 2007. The placebo effect and the theory of the mind. *Pain Practice* 7: 1–3.

Campbell, A. 2009. Hidden assumptions and the placebo effect. *Acupuncture in Medicine* 27: 68–69.

Carel, H. 2008. *Illness*. Stocksfield: Acumen.

Carlino, E., E. Frisaldi, and F. Benedetti. 2014. Pain and the context. *Nature Reviews Rheumatology* 10: 348–355.

Cassell, E.J. 2004. *The nature of suffering and the goals of medicine*. Oxford: Oxford University Press.

Cavanaugh, J. 2015. Performativity. Oxford bibliographies. http://www.oxfordbibliographies.com/view/document/obo-9780199766567/obo-9780199766567-0114.xml. Accessed 1 Jun 2018.

Chan, E.Y., and S.J. Maglio. 2019. Coffee cues elevate arousal and reduce level of construal. *Consciousness and Cognition* 70: 57–69.

Cizza, G., P. Piaggi, K.I. Rother, G. Csako. 2014. For the sleep extension study group. Hawthorne effect with transient behavioral and biochemical changes in a randomized controlled sleep extension trial of chronically short-sleeping obese adults: Implications for the design and interpretation of clinical studies. *PLoS ONE* 9: e104176.

Coan, J.A., H.S. Schaefer, R.J. Davidson. 2006. Lending a hand. Social regulation of the neural response to threat. *Psychological Science* 17: 1032–1039.

Cocco, G. 2009. Erectile dysfunction after therapy with metoprolol: The Hawthorne effect. *Cardiology* 112: 174–177.

Cohen, S. 2014. The nocebo effect of informed consent. *Bioethics* 28: 147–154.

Colagiuri, B., L.A. Schenk, M.D. Kessler, et al. 2015. The placebo effect: From concepts to genes. *Neuroscience* 307: 171–190.

Colloca, L. 2014. Emotional modulation of placebo analgesia. *Pain* 155: 651.

Colloca, L. 2015. Placebo effects in infants, toddlers, and parents (letter). *JAMA Pediatrics* 169: 504–505.

Colloca, L. 2017. Tell me the truth and I will not be harmed: Informed consents and nocebo effects. *American Journal of Bioethics* 17 (6): 46–48.

Colloca, L., and F. Benedetti. 2006. How prior experience shapes placebo analgesia. *Pain* 124: 126–133.

Colloca, L., and F. Benedetti. 2009. Placebo analgesia induced by social observational learning. *Pain* 144: 28–34.

Colloca, L., and F.G. Miller. 2011. How placebo responses are formed: A learning perspective. *Philosophical Transactions of the Royal Society B* 366: 1859–1869.

Colloca, L., D.S. Pine, M. Ernst, F.G. Miller, and C. Grillon. 2016. Vasopressin boosts placebo analgesic effects in women: A randomized trial. *Biological Psychiatry* 79: 794–802.

Conzemius, M.G., and R.B. Evans. 2012. Caregiver placebo effect for dogs with lameness from osteoarthritis. *JAVMA* 241: 1314–1319.

Crum, A.J., and E.J. Langer. 2007. Mind-set matters: Exercise and the placebo effect. *Psychological Science* 18: 165–171.

Crum, A., and B. Zuckerman. 2017. Changing mindsets to enhance treatment effectiveness. *JAMA* 317: 2063–2064.

Darragh, M., R.J. Booth, N.S. Consedine. 2015. Who responds to placebos? Considering the "placebo personality" via a transactional model. *Psychology, Health & Medicine* 20: 287–295.

Darragh, M., R.J. Booth, and N.S. Consedine. 2014. Investigating the 'placebo personality' outside the pain paradigm. *Journal of Psychosomatic Research* 76: 414–421.

de la Fuente-Fernandez, R., T.J. Ruth, V. Sossi, M. Schulzer, D.B. Calne, and A.J. Stoessl. 2001. Expectation and dopamine release: Mechanism of the placebo effect in Parkinson's disease. *Science* 293: 1164–1166.

Enax, L., V. Krapp, A. Piehl, and B. Weber. 2015. Effects of social sustainability signaling on neural valuation signals and taste-experience of food products. *Frontiers in Behavioral Neuroscience* 9: 247.

Faasse, K., A. Huynh, S. Pearson, et al. 2019. The influence of side effect information framing on nocebo effects. *Annals of Behavioral Medicine* 53: 621–629.

Faria, V., C. Linnman, A. Lebel, and D. Borsook. 2014. Harnessing the placebo effect in pediatric migraine clinic. *Journal of Pediatrics* 165: 659–665.

Feinstein, A.R. 1995. Meta-analysis: Statistical alchemy for the 21st century. *Journal of Clinical Epidemiology* 48: 71–79.

Fortunato, J., J. Wasserman, and D.L. Menkes. 2017. When respecting autonomy is harmful: A clinically useful approach to the nocebo effect. *American Journal of Bioethics* 17 (6): 36–42.

Frank, D.L., L. Khorsid, and J.F. Kiffer. 2010. Biofeedback in medicine: Who, when, why and how? *Mental Health in Family Medicine* 7: 85–91.

Franzen, H. 2001. Study finds placebo effect is fake. Scientific American May 21, 2001. https:// www.scientificamerican.com/article/study-finds-placebo-effec/. Accessed 1 Jun 2018.

Frenkel, O. 2008. A phenomenology of the 'Placebo Effect': Taking meaning from the mind to the body. *Journal of Medicine and Philosophy* 33: 58–79.

Garg, A.G. 2011. Nocebo side-effects in cancer treatment. *Lancet Oncology* 12: 1181–1182.

Geuter, S., L. Koban, and T.D. Wager. 2017. The cognitive neuroscience of placebo effects: Concepts, predictions and physiology. *Annual Review of Neuroscience* 40: 167–188.

Goldstein, P., I. Weissman-Fogel, G. Dumas, and S.G. Shamay-Tsoory. 2018. Brain-to-brain coupling during handholding is associated with pain reduction. *Proceedings of the National Academy of Sciences* 115: E2528–E2537.

Gøtzsche, P. 2000. Why we need a broad perspective on meta-analysis; it may be crucially important for patients. *British Medical Journal* 321: 585–586.

Gracely, R.H., R. Dubner, W.D. Deeter, and P.J. Wolskee. 1985. Clinicians' expectations influence placebo analgesia. *Lancet* 325: 43.

Green, M.W., M.A. Taylor, N.A. Elliman, and O. Rhodes. 2001. Placebo expectancy effects in the relationship between glucose and cognition. *British Journal of Nutrition* 86: 173–179.

Grelotti, D.J., and T.J. Kaptchuk. 2011. Placebo by proxy. *British Medical Journal* 343: d4345.

Greville-Harris, M., and P. Dieppe. 2015. Bad is more powerful than good: The nocebo response in medical consultations. *American Journal of Medicine* 128: 126–129.

Gruen, M.E., D.C. Dorman, and B.D.X. Lascelles. 2017. Caregiver placebo effect in analgesic clinical trials for cats with naturally occurring degenerative joint disease-associated pain. *Veterinary Record* 180: 473.

Gupta, A., D. Thompson, A. Whitehouse, T. Collier, B. Dahlof, M. Poulter, et al. 2017. Adverse events associated with unblinded, but not with blinded, statin therapy in the Anglo-Scandinavian Cardiac Outcomes Trial—Lipid-Lowering Arm (ASCOT-LLA): A randomised double-blind placebo-controlled trial and its non-randomised non-blind extension phase. *Lancet* 389: 2473–2481.

Hahn, R.A. 1997. The nocebo phenomenon: Concept, evidence, and implications for public health. *Preventive Medicine* 26: 607–611.

Hall, K.T., A.J. Lembo, I. Kirsch, D.C. Ziogas, J. Douaiher, K.B. Jensen, et al. 2012. Catechol-O-methyltransferase val158met polymorphism predicts placebo effect in irritable bowel syndrome. *PLoS ONE* 7: e48135.

Hall, K.T., J. Loscalzo, and T.J. Kaptchuk. 2015. Genetics and the placebo effect: The placebome. *Trends in Molecular Medicine* 21: 285–294.

Harrar, V., and C. Spence. 2013. The taste of cutlery: How the taste of food is affected by the weight, size, shape, and colour of the cutlery used to eat it. *Flavour* 2: 21.

Hashish, I., H.K. Hai, W. Harvey, C. Feinmann, and M. Harris. 1988. Reduction of postoperative pain and swelling by ultrasound treatment: A placebo effect. *Pain* 33: 303–311.

Häuser, W., E. Hansen, and P. Enck. 2012. Nocebo phenomena in medicine. *Deutsches Ärzteblatt International* 109: 459–465.

Herrnstein, R.J. 1962. Placebo effect in the rat. *Science* 138: 677–678.

Hoover, D.W., and R. Milich. 1994. Effects of sugar ingestion expectancies on mother-child interactions. *Journal of Abnormal Child Psychology* 22: 501–515.

Hróbjartsson, A., and P.C. Gøtzsche. 2007. Powerful spin in the conclusion of Wampold et al.'s re-analysis of placebo versus no-treatment trials despite similar results as in original review. *Journal of Clinical Psychology* 63: 373–377.

Hróbjartsson, A., and P.C. Gøtzsche. 2010. Placebo interventions for all clinical conditions. *Cochrane Database of Systematic Reviews* (1): CD003974.

Hróbjartsson, A., and P.C. Gøtzsche. 2001. Is the placebo powerless? An analysis of clinical trials comparing placebo with no treatment. *New England Journal of Medicine* 344: 1594–1602.

Humphrey, N. 2002. Great expectations: The evolutionary psychology of faith-healing and the placebo effect. In *Psychology at the turn of the millennium*, vol. 2; *Social, developmental, and clinical perspectives*, 225–46, eds. C. von Hofsten, and L. Bäckman. Hove: Psychology Press.

Humphrey, N. 2005. Placebo effect. In *Oxford companion to the mind*, ed. R. Gregory. Oxford: Oxford University Press.

Hunter, T., F. Siess, and L. Colloca. 2014. Socially induced placebo analgesia: A comparison of a pre-recorded versus live face-to-face observation. *European Journal of Pain* 18: 914–922.

Ikemi, Y., and S. Nakagawa. 1962. A psychosomatic study of contagious dermatitis. *Kyushu Journal of Medical Science* 13: 335–350.

Jaksic, N., B. Aukst-Margetic, and M. Jakovljevic. 2013. Does personality play a relevant role in the placebo effect? *Psychiatria Danubina* 25: 17–23.

Jensen, K.B., P. Petrovic, C.E. Kerr, I. Kirsch, J. Raicek, A. Cheetham, et al. 2014. Sharing pain and relief: Neural correlates of physicians during treatment of patients. *Molecular Psychiatry* 19: 392–398.

Jensen, K.B., T.J. Kaptchuk, X. Chen, I. Kirsch, M. Ingvar, R.L. Gollup, et al. 2015. A neural mechanism for nonconscious activation of conditioned placebo and nocebo responses. *Cerebral Cortex* 25: 3903–3910.

Kaasinen, V., S. Aalto, K. Någren, and J.O. Rinne. 2004. Expectation of caffeine induces dopaminergic respomses in humans. *European Journal of Neuroscience* 19: 2352–2356.

Kam-Hansen, S., M. Jakubowski, J.M. Kelley, I. Kirsch, D.C. Hoaglin, T.J. Kaptchuk et al. 2014. Altered placebo and drug labeling changes the outcome of episodic Migraine attacks. *Science Translational Medicine* 6: 218ra5.

Kaptchuk, T.J., J.M. Kelley, L.A. Conboy, R.B. Davis, C.E. Kerr, E.E. Jacobson, et al. 2008. Components of placebo effect: Randomised controlled trial in patients with irritable bowel syndrome. *British Medical Journal* 336: 999–1003.

Kelley, J.M., A.J. Lembo, J.S. Ablon, J.J. Villanueva, L.A. Conboy, R. Levy, et al. 2009. Patient and practitioner influences on the placebo effect in irritable bowel syndrome. *Psychosomatic Medicine* 71: 789–797.

Kennedy, W. 1961. The nocebo reaction. medical. *World* 91: 203–205.

Kessner, S., C. Sprenger, N. Wrobel, K. Wiech, and U. Bingel. 2013. Effect of oxytocin on placebo analgesia: A randomized study. *JAMA* 310: 1733–1735.

Kienle, G.S., and H. Kiene. 1997. The powerful placebo effect: Fact or fiction? *Journal of Clinical Epidemiology* 50: 1311–1318.

Kirchhof, J., L. Petrakova, A. Brinkhoff, et al. 2018. Learned immunosuppressive placebo responses in remal transplant patients. *Proceedings of the National Academy of Sciences* 115: 4223–4227.

Kolata, G. 1998. Scientific myths that are too good to die. *NY Times,* Dec 6.

Lasagna, L., F. Mosteller, J.M. von Felsinger, and H.K. Beecher. 1954. A study of the placebo response. *American Journal of Medicine* 16: 770–779.

Lehner, S.R., C. Rutte, and M. Taborsky. 2011. Rats benefit from winner and loser effects. *Ethology* 117: 949–960.

Levitt, S.D., and J.A. List. 2011. Was there really a Hawthorne effect at the Hawthorne plant? An analysis of the original illumination experiments. *American Economic Journal: Applied Economics* 3: 224–238.

Linde, K., N. Clausius, G. Ramirez, D. Melchart, F. Eitel, L. Hedges, et al. 1997. Are the clinical effects of homoeopathy placebo effects? A meta-analysis of placebo-controlled trials. *Lancet* 350: 834–843.

MacKenzie, J.N. 1886. The production of so-called "Rose-cold" by means of artificial rose. *The American Journal of the Medical Sciences* 91: 45–57.

Mann, C. 1990. Meta-analysis in the breech. *Science* 249: 476–479.

McCarney, R., J. Warner, S. Iliffe, R. van Haselen, M. Griffin, and P. Fisher. 2007. The Hawthorne effect: A randomised, controlled trial. *BMC Medical Research Methodology* 7: 30.

Meiselman, H.L., J.L. Johnson, W. Reeve, and J.E. Crouch. 2000. Demonstrations of the influence of the eating environment on food acceptance. *Appetite* 35: 231–237.

Merleau-Ponty, M. 1962. *Phenomenology of perception*. London: Routledge.

Miller, F.G. 2001. Is the placebo powerless? *The New England Journal of Medicine* 345: 1277.

Miller, F.G., and L. Colloca. 2010. Semiotics and the placebo effect. *Perspectives in Biology and Medicine* 53: 509–516.

Miller, F.G., and L. Colloca. 2011. The placebo phenomenon and medical ethics: Rethinking the relationship between informed consent and risk–benefit assessment. *Theoretical Medicine and Bioethics* 32: 229–243.

Mitsikostas, D.D., L. Mantonakis, and N. Charalakis. 2014. Nocebo in clinical trials for depression: A meta-analysis. *Psychiatry Research* 215: 82–86.

Munana, K.R., D. Zhang, and E.E. Patterson. 2010. Placebo effect in canine epilepsy trials. *Journal of Veterinary Internal Medicine* 24: 166–170.

Nichols, J.T.G. 1893. The misuses of drugs in modem practice. *Massachusetts Medical Society, Medical Communications*. 16: 3–46.

Nichter, M. 2008. Coming to our senses: Appreciating the sensorial in medical anthropology. *Transcultural Psychiatry* 45: 163–197.

Nordström, C. 1998. Terror warfare and the medicine of peace. *Medical Anthropological Quarterly* 12: 103–121.

Oh, V. 1994. The placebo effect: Can we use it better? *British Medical Journal* 309: 69.

Oken, B.S., K. Flegal, D. Zajdel, S. Kishiyama, M. Haas, and D. Peters. 2008. Expectancy effect: Impact of pill administration on cognitive performance in healthy seniors. *Journal of Clinical and Experimental Neuropsychology* 30: 7–17.

Papakostas, Y.G., and M.D. Daras. 2001. Placebos, placebo effect, and the response to the healing situation: The evolution of a concept. *Epilepsia* 42: 1614–1625.

Parellada, M., C. Moreno, M. Moreno, A. Espliego, E. de Portugal, and C. Arango. 2012. Placebo effect in child and adolescent psychiatric trials. *European Neuropsychopharmacology* 22: 787–799.

Parker, S., M. Garry, G.O. Einstein, and M.A. McDaniel. 2011. A sham drug improves a demanding prospective memory task. *Memory* 19: 606–612.

Parsons, H.M. 1974. What happened at Hawthorne? *Science* 183: 922–932.

Paul, I.M., J.S. Beiler, J.R. Vallati, L.M. Duda, and T.S. King. 2014. Placebo effect in the treatment of acute cough in infants and toddlers. A randomized clinical trial. *JAMA Pediatrics* 168 (12): 1107–1113.

Pavlov, I. 1927. *Conditioned reflexes*. Humphrey Milford: Oxford University Press.

Pecina, M., A.S. Bohnert, M. Sikora, E.T. Avery, S.A. Langenecker, B.J. Mickey, et al. 2015. Association between placebo-activated neural systems and antidepressant responses neurochemistry of placebo effects in major depression. *JAMA Psychiatry* 72: 1087–1094.

Peciña, M., H. Azhar, T.M. Love, T. Lu, B.L. Fredrickson, C.S. Stohler, and J.K. Zubieta. 2013. Personality trait predictors of placebo analgesia and neurobiological correlates. *Neuropsychopharmacology* 38: 639–646.

Pedro-Botet, J., and J. Rubiés-Prat. 2017. Statin-associated muscle symptoms: Beware of the nocebo effect. *Lancet* 389: 2445–2446.

Peerdeman, K.J., A.I. van Laarhoven, S.M. Keij, L. Vase, M.M. Rovers, M.L. Peters, et al. 2016. Relieving patients' pain with expectation interventions: A meta-analysis. *Pain* 157: 1179–1191.

Peirce, C.S. 1998. *The essential Peirce*, vol. 2, eds. Peirce edition Project. Bloomington: Indiana University Press.

Peiris, N., M. Blasini, T. Wright, and L. Colloca. 2018. The placebo phenomenon—A narrow focus on psychological models. *Perspectives in Biology and Medicine* 61: 388–400.

Petersen, G.L., N. Brix Finnerup, L. Colloca, M. Amanzio, D.D. Price, T. Staehelin Jensen, et al. 2014. The magnitude of nocebo effects in pain: A meta-analysis. *Pain* 155: 1426–1434.

Petrovic, P., E. Kalso, K.M. Petersson, and M. Ingvar. 2002. Placebo and opioid analgesia—Imaging a shared neuronal network. *Science* 295: 1737–1740.

Pogge, R.C. 1963. The toxic placebo. Part I. Side and toxic effects reported during the administration of placebo medicine. *Medical Times* 91(8): 773–78.

Price, D.D., D.G. Finniss, and F. Benedetti. 2008. A comprehensive review of the placebo effect: Recent advances and current thought. *Annual Review of Psychology* 59: 565–590.

Puustinen R. 2011. Is it Psychosomatic?—An inquiry into the nature and role of medical concepts. Durham theses, Durham University. http://etheses.dur.ac.uk/657/.

Reinoso Carvalho, F., C. Velasco, R. van Ee, Y. Leboeuf, and C. Spence. 2016. Music influences hedonic and taste ratings in beer. *Frontiers in Psychology* 7: 636.

Rogev, E., and G. Pillar. 2013. Placebo for a single night improves sleep in patients with objective insomnia. *Israel Medical Association Journal* 15: 434–438.

Rosenthal, R., and K.L. Fode. 1963. The effect of experimenter bias on performance of the albino rat. *Behavioral Science* 8: 183–189.

Ross, R., C.M. Gray, and J.M.R. Gill. 2015. Effects of an injected placebo on endurance running performance. *Medicine and Science in Sports and Exercise* 47: 1672–1681.

Ruan, X., and A.D. Kaye. 2016. Nocebo effect of informed consent in interventional procedures. *Clinical Journal of Pain* 32: 460–462.

Rütgen, M., E.-M. Seidel, G. Silani, I. Riečanský, A. Hummer, C. Windischberger, et al. 2015. Placebo analgesia and its opioidergic regulation suggest that empathy for pain is grounded in self pain. *Proceedings of the National Academy of Sciences of the United States of America* 112: E5638–46.

Rutte, C., M. Taborsky, and M.W.G. Brinkhof. 2006. What sets the odds of winning and losing? *Trends in Ecology & Evolution* 21: 16–21.

Sacks O. 2007. The abyss. *The New Yorker,* Sept 2. http://www.newyorker.com/magazine/2007/09/24/the-abyss.

Sandler, A.D., and J.W. Bodfish. 2008. Open-label use of placebos in the treatment of ADHD: A pilot study. *Child Care and Health Development* 34: 104–110.

Schienle, A., S. Übel, and W. Scharmüller. 2014. Placebo treatment can alter primary visual cortex activity and connectivity. *Neuroscience* 263: 125–129.

Schwarz K.A., R. Pfister. 2016. Scientific psychology in the 18th century: A historical rediscovery. *Perspectives in Psychological Science* 11: 399–407.

Scott, D.J., C.S. Stohler, C.M. Egnatuk, H. Wang, R.A. Koeppe, and J.K. Zubieta. 2007. Individual differences in reward responding explain placebo-induced expectations and effects. *Neuron* 55: 325–336.

Shiv, B., Z. Carmon, and D. Ariely. 2005. Placebo effects of marketing actions: Consumer may get what they pay for. *Journal of Marketing Research* 42: 383–393.

Shorter, E. 2011. A brief history of placebos and clinical trials in psychiatry. *The Canadian Journal of Psychiatry* 56: 193–197.

Simmons, K., R. Ortiz, J. Kossowsky, P. Krummenacher, C. Grillon, D. Pine, and L. Colloca. 2014. Pain and placebo in pediatrics: A comprehensive review of laboratory and clinical findings. *Pain* 155: 2229–2235.

Slater, R., L. Cornelissen, L. Fabrizi, D. Patten, J. Yoxen, A. Worley, et al. 2010. Oral sucrose as an analgesic drug for procedural pain in newborn infants: A randomised controlled trial. *Lancet* 376: 1225–1232.

Spence, C. 2015. Multisensory flavor perception. *Cell* 161: 24–35.

Steinkopf, L. 2015. The signaling theory of symptoms: An evolutionary explanation of the placebo effect. *Evolutionary Psychology* 13 (3): 1–12.

Stockhorst, U., H.-J. Steingrueber, and W.A. Scherbaum. 2000. Classically conditioned responses following repeated insulin and glucose administration in humans. *Behavioural Brain Research* 110: 143–159.

Stoessl, A., R. dela Fuente-Fernandez. 2004. Willing oneself better on placebo—Effective in its own right. *Lancet* 364: 227–228.

Szawarski, Z. 2004. The concept of placebo. *Science and Engineering Ethics* 10: 57–64.

Tates, K., L. Meeuwesen. 2001. Doctor–parent–child communication. A (re)view of the literature. *Social Science & Medicine* 52: 839–51.

Tausk, F., R. Ader, and N. Duffy. 2013. The placebo effect: Why we should care. *Clinics in Dermatology* 31: 86–91.

Tavel, M.E. 2014. The placebo effect: The good, the bad, and the ugly. *The American Journal of Medicine* 127: 484–488.

Thompson, J.J., C. Ritenbaugh, and M. Nichter. 2009. Reconsidering the placebo response from a broad anthropological perspective. *Culture, Medicine and Psychiatry* 33: 112–152.

Tracey, I. 2010. Getting the pain you expect: Mechanisms of placebo, nocebo and reappraisal effects in humans. *Nature Medicine* 16: 1277–1283.

Turnwald, B.P., J.P. Goyer, D.Z. Boles, et al. 2019. Learning one's genetic risk changes physiology independent of actual genetic risk. *Nature Human Behaviour* 3: 48–56.

Ueberwasser, F. 1787. Anweisungen zum regelmäßigen Studium der Empirischen Psychologie für die Candidaten der Philosophie zu Münster. Münster: Friedrich Christian Theißing.

Vachon-Presseau, E., S.E. Berger, T.B. Abdullaah, et al. 2018. Brain and psychological determinants of placebo pill response in chronic pain patients. *Nature Communications* 9: 3397.

Valentini, E., M. Martini, M. Lee, S.M. Aglioti, and G.D. Iannetti. 2014. Seeing facial expressions enhances placebo analgesia. *Pain* 155: 666–673.

Vogt, H., E. Ulvestad, T.E. Eriksen, and L. Getz. 2014. Getting personal: Can systems medicine integrate scientific and humanistic conceptions of the patient? *Journal of Evaluation in Clinical Practice* 20: 942–952.

Walker, L.S., S.E. Williams, C.A. Smith, J. Garber, D.A. Van Slyje, and T.A. Lipani. 2006. Parent attention versus distraction: Impact on symptom complaints by children with and without chronic functional abdominal pain. *Pain* 122: 43–52.

Wampold, B.E., T. Minami, S. Callen Tierney, T.W. Baskin, and K.S. Bhati. 2005. The placebo is powerful: Estimating placebo effects in medicine and psychotherapy from randomized clinical trials. *Journal of Clinical Psychology* 61: 835–854.

Wearing, D. 2005. *Forever today*. Reading: Corgi Books.

Weger, U.W., and S. Loughnan. 2013. Mobilizing unused resources: Using the placebo concept to enhance cognitive performance. *Quarterly Journal of Experimental Psychology (Hove)* 66: 23–28.

Weimer, K., M.D. Gulewitsch, A.A. Schlarb, J. Schwille-Kiuntke, S. Klosterhalfen, and P. Enck. 2013. Placebo effects in children: A review. *Pediatric Research* 74: 96–102.

Wells, R.E., and T.J. Kaptchuk. 2012. To tell the truth, the whole truth, may do patients harm: The problem of the nocebo effect for informed consent. *American Journal of Bioethics* 12 (3): 22–29.

Whalley, B., and M.E. Hyland. 2013. Placebo by proxy: The effect of parents' beliefs on therapy for children's temper tantrums. *Journal of Behavioral Medicine* 36: 341–346.

Widdershoven, G., G. Meynen, and S. Metselaar. 2017. Dealing with the nocebo effect: Taking physician-patient interaction seriously. *American Journal of Bioethics* 17 (6): 48–50.

Wiech, K. 2016. Deconstructing the sensation of pain: The influence of cognitive processes on pain perception. *Science* 354: 584–587.

Zunhammer, M., U. Bingel, and T.D. Wager. 2018. Placebo effects on the neurologic pain signature. *JAMA Neurology* 75: 1321–1330.

Chapter 4
Clinical Use of Placebos

Deliberate use of placebos, defined as treatments without any effect as such, is usually considered unethical. The Council on Ethical and Judicial Affairs of the American Medical Association, for example, states that it "may undermine trust, compromise the patient–physician relationship, and result in medical harm to the patient" (AMA 2018).

Physicians' attitudes on the use of placebos vary broadly, however, reflecting the different meanings given to the concept. Examples of two extreme positions are those of Brown (1998): "As physicians, we should respect the benefits of placebos—their safety, effectiveness and low cost—and bring the full advantage of these benefits into our everyday practices", and Hróbjartsson (2008): "Clinical placebo interventions are unethical, unnecessary, and unprofessional."

4.1 Are Placebos Used in Clinical Practice?

During the past three decades, several studies have come to the conclusion that the deliberate use of placebos is a common and widely accepted practice. A closer look at the studies shows, however, that the placebo concept is understood in several different ways and that the crucial difference between placebo use and placebo effect is often not understood (Louhiala 2012). Three examples of these studies demonstrate that it is by far from clear what can be concluded from their results.

A study in Denmark found that 86% of the general practitioners included in the sample reported having used a placebo intervention at least once during the previous year (Hróbjartsson and Norup 2003). The researchers deliberately used a narrow definition for the placebo, as an intervention not considered to have any 'specific' effect on the condition treated but with a possible 'unspecific' effect. In the paper, the authors did not give a definition for these concepts but, instead, wrote: "Whereas this distinction is relatively clear in connection with pharmacological interventions,

© Springer Nature Switzerland AG 2020
P. Louhiala, *Placebo Effects: The Meaning of Care in Medicine*,
International Library of Ethics, Law, and the New Medicine 81,
https://doi.org/10.1007/978-3-030-27329-3_4

it is less obvious in connection with nonpharmacological treatments such as physio-therapy." As we saw in Sect. 2.2, the concepts 'specific' and 'unspecific' are vague and open to many interpretations.

According to the authors, one typical 'placebo intervention' among Danish physi-cians concerned antibiotics for viral infections. This use of the term may sound sim-ple, but the interpretation is not obvious, as the authors also note in their discussion. Respiratory infections are viral by origin in most cases, but, particularly in general practice, laboratory tests are hardly ever made for confirmation. In prolonged or atypical cases, physicians often prescribe antibiotics although no firm evidence of a bacterial origin exists. Furthermore, many patients have both viruses and bacteria, and it is not even possible to determine for sure whether antibiotics are necessary.

The main finding of a questionnaire survey in Israel was that 60% of the respon-dents (53% of the physicians and 71% of the nurses who were sent the questionnaire) reported using a placebo (Nizan and Lichtenberg 2004). Most of the users found placebos generally or occasionally effective. Again, the results are difficult or impos-sible to interpret because the authors did not specify what they meant by placebo. Because the key concept was not defined, the respondents may have understood placebo in at least four different ways (Louhiala 2009).

First, placebo may have meant deliberate deception through the administration of an inert substance. Second, it may have meant giving an inert substance openly. (In fact, 29% of the respondents thought that patients should be informed that they were receiving a placebo). Third, some may have thought of a situation in which the doctor or nurse believes that the drug works even though there is no supporting scientific evidence. Fourth, some respondents may have thought more about the placebo effect than the nature of the substance given.

The respondents' description of the circumstances of the use of a placebo demon-strates the variation in their understanding. For example, paracetamol and vitamin C were mentioned as examples of placebos given. Although the conditions in which the placebo were used were not reported, it is obvious that neither substance can be labelled inert.

In the abstract of the paper, the authors concluded that "used wisely, placebos might have a legitimate place in therapeutics". This conclusion is unjustified for two reasons. First, on the basis of the survey it is far from clear *what* might have a legitimate place in therapeutics. Second, the authors report the results of an empirical descriptive study and do not scrutinise the normative arguments for or against the clinical use of a placebo.

According to a questionnaire survey among internists in Chicago, 45% of the respondents had used placebos in clinical practice (Sherman and Hickner 2008). The researchers provided several alternatives for the definition of placebo. The respon-dents could give their own definition or choose between the following alternatives, one of which was the same as in the Danish study: (1) an intervention that is not expected to have an effect through a known physiologic mechanism; (2) an inter-vention not considered to have a 'specific' effect on the condition treated, but with a possible 'unspecific' effect; and (3) an intervention that is inert or innocuous.

The definitions demonstrate the conceptual ambiguity. The first definition includes the hint of a possible psychological mechanism, and the second contains the problematic terms 'specific' and 'unspecific'. The third definition is narrow and refers to the characteristics of the intervention only.

In this study, too, the description of the treatments that were understood as placebos demonstrates the multiple meanings of the term. Antibiotics (for viral or other non-bacterial diagnoses), vitamins, ibuprofen, herbal supplements, and prepared placebo tablets are examples of placebo therapies mentioned by the respondents.

The authors briefly refer to the discussion on the ethics of deliberate placebo use. However, they confuse the use of placebo and placebo effect when they write: "In the broader ethics literature, some commentators on informed consent and nondeceptive therapeutics caution against the use of placebos in medical practice. Others propose that the placebo effect can be harnessed in various therapeutic contexts that do not pose ethical dilemmas."

In summary, the results of these surveys demonstrate the confusion around the concepts placebo and placebo effect. The conclusion that placebos are commonly used in clinical practice is thus not meaningful.

4.2   Impure Placebo—Conceptual Confusion Revisited

In the empirical studies 'placebo' was either explicitly or implicitly understood as an upper-level category that is divided into two lower-level categories, 'pure' and 'impure' placebos (Louhiala et al. 2015). In the surveys these two categories were taken for granted; yet their problematic nature was also occasionally acknowledged:

> What is considered to be an impure placebo varies considerably among studies and it is unclear and subjective when an intervention is a placebo or an active or effective intervention. Surveys investigating definitional aspects reveal considerable disagreement regarding whether defined interventions should be considered (impure) placebos or not… With this lack of clarity it is doubtful whether the evaluation of something such as 'the prevalence of the use of impure placebos' makes sense. (Fässler et al. 2010)

The authors conclude by noting that "the academic concept of an impure placebo might inappropriately reflect the complex situations and motivations in which health care professionals apply interventions which are not backed up by scientific evidence". Acknowledging the problems in using the impure placebo concept, however, did not prevent the same authors from using that notion in a subsequent article, "Widespread use of pure and impure placebo interventions by GPs in Germany" (Meissner et al. 2012).

I will argue below that the impure placebo concept is not just scientifically unsound but may also be actively harmful. It causes confusion for both the scientific community and the general public in particular because the term 'placebo' carries a negative connotation in the context of clinical medicine.

The origins of the concept

The earliest division of the placebo concept into the subconcepts of pure and impure placebo that I have been able to trace was in the second volume of the Cornell Conferences on Therapy in the 1940s, which included a lengthy discussion titled "The use of placebos in therapy" (Gold et al. 1947). In that discussion, Dr. Eugene DuBois divided placebos into three classes. The first is pure placebos, the traditional bread pills or lactose tablets with no significant physiological effects. The second is impure placebos, which are "adulterated with a drug that might have some pharmacological action, such as a tincture of gentian or a very small dose of nux vomica. That is the adulterated placebo, the false placebo, the bastard placebo, you might call it" (ibid., p. 4). The third group of placebos in DuBois's classification is "the universal pleasing element which accompanies every prescription". DuBois's first and third classes have traditionally been the source of confusion about placebos (Louhiala 2012). The third class refers to practical aspects of good doctoring, i.e., the placebo effect generated by physicians.

After DuBois's published discussion, the impure placebo concept has been occasionally mentioned in the medical literature (Leslie 1954; Shapiro and Morris 1978). Nevertheless, the concept has remained unfamiliar in wider scientific circles. Recently, the impure placebo concept was used in surveys on how extensively physicians use placebos, and this attention has made the concept currently important. To illustrate the problems of the impure placebo, I will first analyse three definitions of the concept, as presented in the literature. I will then discuss examples of impure placebos that are explicitly mentioned in the surveys.

Definitions of impure placebo

Definitions of impure placebo given in the surveys overlap significantly, and therefore, I will describe only three definitions from studies addressing Swiss general practitioners and paediatricians (Fässler et al. 2009), Swiss general practitioners (Fent et al. 2011), and British general practitioners (Howick et al. 2013). Fässler et al. (2009) define impure placebos as "substances or methods which do have a *known* pharmacological or physical activity but which cannot be expected to have any *direct* therapeutic effects for the respective disease and in the chosen dosage" [italics added]. Fent et al. (2011) consider impure placebos to have pharmacological effects, "but the effect on the specific disease the substance is prescribed for has *not been proven* or *is uncertain*" [italics added]. Howick et al. (2013) state that impure placebos are "substances, interventions or 'therapeutic' methods which have *known* pharmacological, clinical or physical value for some ailments but *lack specific* therapeutic effects or value for the condition for which they have been prescribed" [italics added].

There are several problems in these definitions. First, it is important to distinguish between the following two questions: (1) does an intervention have an effect? and (2) what is the mechanism of the effect? The distinction between intervention and mechanism is not made in the above definitions. There are effective interventions for which we do not know the mechanism, yet they are effective. Similarly, there are known mechanisms that do not correspond to clinically relevant effects.

As we saw in Sect. 2.2, terms such as 'specific', 'unspecific', 'direct' or 'indirect' are commonly used in the placebo literature, but remain vague and unhelpful. If an intervention does not have a specific or direct effect, referring to a nonspecific or indirect effect is just another way of saying that the mechanisms of the intervention are not yet known. However, this does not mean that the intervention is ineffective. Furthermore, especially within the field of general practice, which is where the surveys on impure placebo have mainly been carried out, the symptoms and worries that the patients present to their physicians are often nonspecific in the medical sense. Often, there are no specific treatments in the same sense as those that exist for specialist care with more severe and well-defined diseases. But I do not see any justification for labelling the alleviation of patients' symptoms and worries 'impure placebo'.

Second, if the effect of a treatment 'has not been proven', it does not imply that there are no effects. "Absence of evidence is not evidence of absence" is the message of an important paper in the British Medical Journal (Altman and Bland 1995). Smoking, for example, was harmful before the harm had been undisputedly proven in large cohort studies. Furthermore, 'has not been proven' is a question which is often limited by resources. Pharmaceutical companies have extensive resources for randomised controlled trials to test their own patented drugs, but similar resources are not available for most non-patentable treatments. There is no basis for believing that all untested treatments are ineffective.

Third, the meaning of 'uncertain' is not clear in the definition by Fent et al. (2011). They may, for example, refer to lack of consensus in the medical community. In this meaning, certainty and uncertainty do not represent a dichotomy but refer to extreme points on a continuum. Furthermore, in clinical practice, many or most treatments are uncertain in the statistical sense. For example, if a treatment helps 50% of the participants in a randomised trial, the clinician does not know to which half his or her patient belongs, even when it is reasonable to use the treatment. Average benefit does not guarantee that a treatment works for a particular patient. Inconsistencies between definitions create a situation in which researchers do not measure the same phenomenon. As Meissner et al. (2012) write, "A major problem in any survey on placebo use is the concept of impure placebo. The decision whether an intervention is still an active treatment or a placebo depends on personal attitudes and situational factors" (ibid., p. 81). As long as there is no common agreement on how impure placebos can be defined unambiguously, the scientific value of surveys on their use is low. By analogy, if different psychiatrists were to each have their own definition for 'depression', studies addressing the prevalence of depression in different countries could not be compared.

Common examples of 'impure placebos'

Apart from the conceptual problems related to the definition of the term 'impure placebo', there are also problems with classifying actual treatments into that category. Below I comment shortly on treatments that have been mentioned as examples of 'impure placebos in at least two surveys (for more examples, see Louhiala et al. 2015).

Antibiotics for suspected viral infections. Most infections of the upper respiratory tract are viral in origin, and therefore, antibiotics are not beneficial for most of these infections. However, in a clinical setting, it is usually not possible to be fully certain that a particular infection is not caused, at least partly, by bacteria. Kaiser et al. (1996) found that antibiotics were clinically beneficial for a subgroup of patients with upper respiratory infections whose nasopharyngeal secretions contained Haemophilus influenzae, Moraxella catarrhalis, or Streptococcus pneumoniae. In practice, physicians prescribe antibiotics on the basis of the nature, severity, and duration of the symptoms of individual patients (Louhiala 2009; Young et al. 2008). It is not reasonable to classify all antibiotic use for 'suspected viral infections', a diffuse category, as impure placebo treatment.

Non-essential physical examinations and non-essential technical examinations of a patient (blood tests, X-rays). It is not possible to be certain which physical or technical examinations are diagnostically essential for an individual patient. All experienced physicians recall cases in which an examination that originally seemed non-essential unexpectedly revealed important findings and vice versa. There is no sharp line between obviously essential and obviously unsound examinations. Instead, there is a wide region in which the physician must make subjective decisions, and different physicians make different decisions. 'Non-essential examinations' is not a reasonable category for impure placebos.

Peppermint pills for pharyngitis. To use an analogy, pain killers do not kill viruses, and in that respect, they do not influence the aetiology of viral pharyngitis. However, pain killers improve the quality of life of patients who have sore throats, and improved quality of life is one of the most important goals of medicine. Similarly, peppermint pills may alleviate the pain of a sore throat or cough, and there is a biological rationale for this effect. The pills increase the production of saliva, and the saliva increase may protect the mucosa of the pharynx from airflow and aid the drainage of mucus. There are no randomised trials examining the effect of peppermint pills on sore throat, but the experience of numerous patients, including those of the author, indicates that peppermint pills are useful in alleviating the symptoms of pharyngitis. It is unlikely that pharmaceutical companies will become interested in carrying out trials on peppermint pills, and it is also unlikely that public health authorities will prioritize funding for such trials. However, a lack of published trials does not imply that peppermint pills are ineffective in improving the quality of life of pharyngitis patients.

Phytotherapeutics/herbal supplements. This is a broad category, and it is not meaningful to pool all such treatments into a single group and classify them as impure placebos, as was done in the two surveys. Although I am sceptical of most treatments falling into this category, there is evidence that some phytotherapeutic preparations are effective. St. John's wort extracts, for example, have been beneficial in the treatment of mild to moderate depression in several placebo-controlled trials.

Positive suggestions. Physicians often assure their patients with such statements as, 'You will certainly feel better in a few days' or 'I'm sure this will help you'. Such utterances are an essential part of good doctoring in any clinical setting all over the world. It is not reasonable to argue that such assertions cannot be expected to have any direct therapeutic effect for the respective disease and in the chosen dosage. To classify good doctor-patient relationships as impure placebos does not make sense.

Sedatives. This is a broad category that includes, for example, benzodiazepines. Sedatives are used for a large number of medical indications, and it is unreasonable to dichotomise their clinical use into indicated use and use as impure placebo on the basis that they either have or lack 'specific therapeutic effects or value for the condition for which they have been prescribed'.

Sub-clinical doses of otherwise effective therapies. Individual patients respond to medication differently because they metabolise drugs differently. A subclinical dose for one patient may be clinically effective for another, and 'normal' doses may be too low for some patients. Particularly in symptomatic treatment, it is common to start with a low dose and slowly increase the dose until the patient responds or has unwanted side effects (Novella 2013). The use of drugs in low doses cannot be labelled categorically as an 'impure placebo'.

Vitamin infusions for cancer. Using vitamin C to treat cancer patients is not just a question of whether a high concentration of vitamin C might kill cancer cells. There is evidence that a substantial proportion of hospital patients have a vitamin C deficiency (Holley et al. 2011; Raynaud-Simonn et al. 2010). Therefore, vitamin C may be beneficial for some hospital patients with or without cancer (Zhang et al. 2011; Wang et al. 2013). There are also relevant case reports of vitamin C infusions being used to treat cancer patients (Padayatty et al. 2006). There is no basis to classify vitamin C treatments of cancer patients categorically as impure placebos.

Vitamins without approved indications. Lack of vitamins was identified during the first part of the twentieth century as the explanation for diseases that we nowadays classify as deficiency diseases. For most vitamins, the approved indication is to treat deficiency, but this does not mean that they cannot have other effects. For example, systematic reviews have shown that vitamin C shortens the duration of colds (Hemilä and Chalker 2013), reduces the decline in forced expiratory volume (FEV1) caused by exercise (Hemilä 2013), and reduces blood pressure (Juraschek et al. 2012). There is no basis to classify all use of vitamins for purposes other than treating deficiency diseases as an impure placebo, even though the clinical role of vitamins in non-deficiency diseases is still a controversial issue.

The preceding treatments were listed as examples of 'impure placebos' in at least two surveys that address the use of placebos by physicians. I hope to have demonstrated that each example is unsound from the point of view of both medical science and clinical practice.

Methodological problems in surveys on the use of impure placebos

Surveys on impure placebos have queried primary care physicians on their use of such placebos by sending questionnaires to a number of practising physicians. In addition to the issues related to the concept of impure placebo as such, there are other methodological problems with the surveys.

First, it is unclear how the responding physicians interpreted the definitions of impure placebos when they considered the examples of impure placebos in the list contained in the questionnaires. Some physicians may have counted all cases in which they prescribed a substance or method on the list, while others may have counted only those cases in which they prescribed impure placebos to please the patient. Thus, there can be large variation in how the respondents interpreted the survey questions.

Second, it is unclear how the physicians surveyed interpreted 'prescription' or 'use' of impure placebos. At least the following interpretations are possible: (1) writing a formal prescription for an impure placebo without informing the patient that the physician believed the impure placebo to be pharmacologically ineffective in this particular case, (2) verbally suggesting that the patient might test a treatment the physician considered ineffective, and (3) allowing the patient to continue the use of alternative medicine treatments that the patient had initiated himself or herself although the physician considers them useless. These situations differ to an important degree, but they have not been addressed in the surveys.

The impure placebo concept may even be harmful

The impure placebo concept is not just meaningless, but may be harmful, for the following reasons. First, ambiguous concepts lead to ambiguous thinking. For example, conclusions like "placebos are commonly used in UK primary care" (Howick et al. 2013) and "prescribing placebo treatments seems to be common" (Tilburt et al. 2008) are meaningless because they are based on combining the use of both so-called pure and impure placebos. Most readers assume that 'placebo' in such a context refers to pure placebo, and it is unlikely that readers understand that physicians' positive assertions, peppermint pills, sedatives, administering antibiotics for the common cold, etc. are being counted as placebos.

Second, published conclusions distort the impression of medicine in the eyes of the general public. Media coverage of the study by Jon Tilburt et al. (2008) included the headline "U.S. Doctors Regularly Prescribe Real Drugs as Placebo Treatments, Study Claims" (ScienceDaily 2008). In a similar way, the coverage of the Howick et al. (2013) study included the following headlines: "Many UK Doctors Give Useless Drugs, Treatments" (Cheng 2013), and "The Placebo Effect: Doctors Admit Prescribing Unproven Treatments, Unnecessary Tests and Pills with No Active Ingredient" (Philby 2013). Such secondary reporting paints a grossly misleading view of medical practice and may create unjustified mistrust among the general public towards the medical profession.

Third, the impure placebo concept undermines the importance of the physician-patient relationship and the context of care. The associations with the term 'placebo' are negative, since placebos are commonly described with terms such as 'dummy' and 'sham', and the placebo effect with expressions such as 'suggestion'. Therefore, terms including the word 'placebo' carry pejorative connotations regardless of what has been written about their importance, existence, and supposed mechanisms.

Fourth, because the concept has no unambiguous definition, survey research on the use of 'impure placebos' is a waste of scarce resources in medical research.

Conclusions

In my view, positive suggestions in physician-patient relationships and peppermint pills for sore throats are examples of important components of good doctoring. They should not be negatively labelled as forms of placebo ('sham' or 'dummy') treatment. The introduction of the impure placebo concept alongside the traditional pure placebo, which universally is simply called 'placebo', leads to confusion. In particular, when impure placebos are pooled with pure placebos and, together, are called a 'placebo', readers are easily misled because 'placebo' refers to 'pure placebo' in ordinary language. As I have shown, the impure placebo concept is poorly defined. In addition, the examples commonly given for impure placebos are unsound and, in some cases, absurd from the point of view of clinical work. The impure placebo is a useless concept and should not be used in scientific or medical literature. The issues *behind* the concept, e.g., uncertainty in clinical practice and its consequences on treatment decisions, however, deserve serious attention in future research.

4.3   The Ethics of Using Placebos in Clinical Practice

It is widely believed that a beneficial response to placebo treatment requires deception or at least some kind of a "white lie" to the patient. Doctors have, in fact, lied to their patients throughout history. *Mentiris ut medicus* ("lie like a doctor") was a popular proverb in the Middle Ages (Carlino 2005) and in the 17th century Bishop Jeremy Taylor encouraged physicians to lie if it helped the patients (Bok 2002). Lying was, however, to be understood with charity and with honour to the profession. Telling the truth as a moral imperative is, in fact, a fairly recent phenomenon and is closely related to the development of patient autonomy in the Western world since the 1960s. It is worth noting also that withholding the truth—a serious diagnosis, for example—is still common practice in many, maybe even most, parts of the world.

Lying to the patient is, however, ethically problematic, to say the least. For example, if the patient later finds out that the physician has not told the whole truth about the treatment, there may be serious consequences not only for the present physician-patient relationship but also for future relationships with other health-care professionals.

Given the conceptual confusion in defining the concept placebo, it is not surprising that opposite views have been presented also about the acceptability of the use of placebos in clinical practice. At the other extreme, Howick et al. (2013) have proposed further investigations to develop "ethical and cost-effective placebos" and Lichtenberg et al. (2004) have suggested that "in select cases, use of the placebo may even be morally imperative."

A simplified version of the argument in support of the use of placebos can be summarized as follows: "they work in clinical trials, are cheap, and have no side effects." American psychiatrist Walter Brown, for example, defended clinical place-bos this way in an article in Scientific American in 1998 (Brown 1998). Brown himself had done research on psychiatric drugs as well as placebo effects. In the article he reviewed some earlier research and gave credit to the doctor-patient rela-tionship and the healing environment but confused the concepts placebo and placebo effect in the same way as numerous authors before and after him:

> Placebos are usually defined not in terms of what they are but what they are not. They are often described as inactive, but placebo agents are clearly active: they exert influence and can be quite effective in eliciting beneficial responses.

Brown suggested that:

> If physicians can see placebos—like many conventional drugs—as broadly effective thera-pies, whose mechanisms of action are not completely understood and which tend to be more effective for some conditions than others, they can then offer placebos both honestly and as plausible treatment. … The specific placebo chosen should be free of toxicity and should be in keeping with the patient's beliefs and expectations.

Brown's conclusion was that "as physicians, we should respect the benefits of place-bos—their safety, effectiveness and low cost—and bring the full advantage of these benefits into our everyday practices."

The only unproblematic part is the last one: 'they' are cheap, if 'placebo' refers to a practically inert substance. 'They are effective' is trivially false if 'they' refer to inert substances. Otherwise 'they are effective' seems to refer to 'placebo effects' in general and in that case 'they' are the whole context of the treatment and not the placebos as such. 'They are safe' is trivially true about inert substances but false referring to 'placebo effects' (or actually nocebo effects) in a broader sense.

A more elaborated defence of the use of placebos in clinical practice has been presented by Nitzan and Lichtenberg (2004) and Evans and Hungin (2007).

A moral imperative?

After briefly introducing the conceptual problems related to placebos, Nitzan and Lichtenberg (2004) present three clinical cases on the basis of which they propose guidelines for the clinical use of placebos and present their claim about the occasional duty to use placebos.

The paper contains, however, questionable reasoning and erroneous assumptions. First, the operational definition of the placebo concept is problematic. The authors write that they

> … address the ethics of the placebo operationally by asking when it is ethical, in clinical practice, to offer a pill or perform a procedure as an alternative to, or in the absence of,

a standard, proven therapy when the effect, if any, of that pill or procedure is expected to be mediated by psychophysiological mechanisms, such as expectation, relaxation, or conditioned response, or what has elsewhere been termed a 'meaning response'. The pill or procedure would then be considered the placebo; the effect it produces would be the placebo effect.

If, however, a standard, proven therapy exists, on what grounds could it ever be ethically acceptable to offer a placebo as an alternative to it? In the absence of a standard, proven therapy, giving a placebo and telling the patient openly about it can be an option (see below). The key moral issue here is deception.

The authors connect the morality of placebo prescription to a particular world view: "The placebo is a deception only for those who would reduce treatment to a purely biomedical pursuit". This statement is problematic. An inert treatment as such is not deception, but the act of providing the treatment may or may not contain deception.

Only the first of the three cases presented in the paper supports the possible open use of a supposed placebo medication. A 45-year-old man had severe postoperative pain after leg amputation. Opioids did not help and he was offered intramuscular saline. The staff explained that saline had been used as an effective pain killer, and it did help this patient, too. In the second case, the wife of a patient demanded a shot of penicillin for her husband, who had gastroenteritis. The doctor did not agree to give the shot, but the authors concluded, "Had the family continued to demand treatment beyond reassurance, the doctor might have considered giving saline, admitting it was saline, and assuring the patient that he would rapidly recover." The authors do not, however, present any arguments to support their claim. A 'demand beyond reassurance' is a challenge for communication, not a legitimation for an injection of intramuscular saline, even when given openly.

The third case was that of a 32-year-old woman with depression, who, after failed attempts with hypnotherapy, demanded medication. The psychiatrist prescribed imipramine at a starting dose of 25 mg, explaining that effectiveness generally requires 2–4 weeks at a much higher dose. The patient reported remarkable improvement, however, already the next day. The authors thought that imipramine was used here as a placebo. This is not obvious, however. Even a low dose of a tricyclic antidepressant cannot be considered an inert substance. Also, individual sensitivity to drugs varies greatly, and even the first dose might have helped the patient if it had, for example, made her sleep better.

Chocolate sweets rather than placebos?

Evans and Hungin (2007) describe a fictional general practice consultation, involving Mr. Smith, a patient with irritable bowel syndrome, and his general practitioner, Dr. Jones. Their discussion concentrates on a placebo, 'SAS', a specific drug that has been developed for the treatment of irritable bowel syndrome, and 'Swotties', well-known proprietary chocolate sweets and "a harmless and enjoyable substance". An important detail is that, in clinical trials, the effectiveness of SAS and a placebo have been the same.

After a lengthy and sometimes complex chain of arguments, Evans and Hungin end up with two main conclusions:

when both (a) a drug fails to out-perform placebo and (b) the condition in question is a functional illness with no demonstrable underlying pathology, then the action of the drug is not only no better than placebo, and it is also no different from it either.

Furthermore, the authors claim that,

... in the circumstances of the consultation described, it is striking that current governance deems it ethical for a practitioner to prescribe either a drug or a placebo, both of which appear to rely for their effectiveness on a measure of concealment on the part of the doctor, yet deems it unethical for a practitioner openly to prescribe a harmless and enjoyable substance which (in equivalent conditions of transparency and information) is likely to be no less effective than either drug or placebo and is also likely to be better-tolerated and cheaper than the drug.

Dr. Jones also concluded at the end of the paper that "the three preparations on her desk—the pharmaceutical product SAS, the 'official' placebo and the green-candied Swotties, are 'functionally equivalent'".

The fictional consultation is described clearly; there are, however, several problems in the argumentation. First, it cannot be argued that the action of the drug does not differ from that of the placebo if it has failed to outperform a placebo in clinical trials. Even so, the drug as such may have some unwanted side effects, while placebos as such are, by definition, inert.

The authors do not pay attention to the fundamental difference between the two contexts in which placebos are used or can be used—the clinical context and the clinical trial context. Their difference is obvious if we consider the main mechanism of the placebo effect, namely, expectations. In the clinical encounter, the expectations of both the doctor and the patient are usually high. In a clinical trial, the patients are told that after randomisation they may end up in a placebo group; they understand this alternative and give their informed consent. The aim of the clinical encounter is to promote the health of the patient, while the aim of clinical research is to promote medical science and the health of future patients. Sometimes the trial participants are helped and sometimes they are harmed. The results of a clinical trial cannot simply be transferred into clinical practice.

It is not true that "current governance deems it ethical for a practitioner to prescribe either a drug or a placebo, both of which appear to rely for their effectiveness on a measure of concealment on the part of the doctor". While the open prescription of placebos in some conditions is accepted, concealed prescription is not. For example, the American Medical Association (AMA 2018) is explicit in its policy: "Physicians may use placebos for diagnosis or treatment only if the patient is informed of and agrees to its use."

Further, it cannot be said that Swotties are functionally equivalent to a placebo and SAS. While it is true that the latter two have been similarly successful in trials, no trials have used Swotties.

Finally, Evans and Hungin argue that the lack of transparency is "generally supposed to underlie successful use of placebo—namely, to enjoy the benefits of the power of suggestion, it seems that the doctor must conceal from the patient at least some of the pertinent facts about the substance prescribed". This general supposition is, however, not true. Placebos have been successfully used openly in clinical trials (see Sect. 4.5.). A placebo effect can be produced by suggestions, but other mechanisms may be far more important. Past effects of active treatments and cues that signal that an active medication or treatment has been given are some such mechanisms.

4.4   Patient Autonomy and the Use of Placebos

Patient autonomy was introduced as one of the four principles of modern bioethics or health care ethics, when they were introduced in the 1960's and 1970's. Very soon autonomy became the dominant principle, particularly in American bioethics. The expansion of autonomy was welcome and it made the doctor-patient relationship "more open, more adult, more transparent, and more attentive to the patient's values and wishes" (Pellegrino 2009).

A few decades later it became clear, however, that respect for autonomy had overridden other important values, leading to situations in which patients feel that they have been abandoned (Louhiala 2014). It had been forgotten that people are, after all, essentially dependent beings (Nys et al. 2007). Becoming a patient radically alters the character of personal identity when compared with the normal setting and "a radically independent, autonomous person is at best an idealised portrait of a fictional character, part of an elaborate ideological cartoon of Western culture" (Tauber 2005).

According to Tauber, "medical ethics generally, and patient autonomy in particular, filled an ethical lacuna left by the erosion of patient trust, and thus patient autonomy became the sacrosanct principle governing medical ethics" (ibid.). O'Neill argued that "conceptions of individual autonomy cannot provide a sufficient and convincing starting point for bioethics, or even for medical ethics. They may encourage ethically questionable forms of individualism and self-expression and may heighten rather than reduce public mistrust in medicine, science and biotechnology" (O'Neill 2002).

Loewy, in his paper "In Defence of Paternalism", maintained that "the pendulum has swung from rather crass paternalism practised fifty or sixty years ago to an obsession with autonomy which allows patients with questionable autonomy to come to harm" (Loewy 2005). He accused autonomy of becoming hidden paternalism that "abandons patients to their autonomy and makes a relationship which should be mutual and caring, into a cold business transaction" (ibid.). Feder Kittay went as far as asking the questions: "Do we really need the terms paternalism and autonomy? Do they obscure more than they illuminate?" (Feder 2007).

Empirical evidence confirms the critique and shows that many patients do not wish to make their own medical decisions (Schneider 2005). It has been found, for example, that patients prefer that decisions be made principally by their physicians, not themselves, although they very much want to be informed (Ende et al. 1989). For most of their patients, the desire to make decisions declined as they faced more severe illness. Older patients had less desire than younger patients to make decisions and to be informed.

Empirical discoveries in placebo studies also call into question the demarcation between the principles of autonomy and beneficence in considering the ethics of the doctor-patient relationship (Annoni and Miller 2016). These studies demonstrate that "the way in which health professionals communicate, disclose, frame, and contextualize information to patients may modulate symptoms across an array of highly prevalent conditions such as pain, depression, anxiety, insomnia, irritable bowel syndrome, migraine, and Parkinson's disease" (ibid.).

Authorized concealment and deception

Expectations modify the effects of treatments and the same treatment will have different effects depending on how the patient has been informed. Because of this, Alfano (2015a) calls for a "nuanced, empirically informed conception of autonomy." He points out that "this allegedly univocal value can even conflict with itself; for instance, more information (a component of autonomy) sometimes leads to worse decision-making processes (another component of autonomy)" (ibid.). Alfano's solution to the dilemma is *authorized concealment* and modification of it, *authorized deception*.

He explains authorized concealment in the light of an example:

> Suppose a doctor is consulting with a man who seeks treatment for male pattern baldness. She recommends finasteride. As I mentioned earlier, one side effect of finasteride is erectile dysfunction, which is 300% more likely when the patient has been led to expect it. Instead of telling the patient about this side effect, the doctor might say, 'As with any hormone therapy, this treatment carries the risk of side effects. Research suggests that if I tell you what those side effects are, you'll be much more likely to experience them. So, what I'd like to do is ask your permission not to mention these side effects. I'll debrief you about them after the treatment is over.' (Alfano 2015a)

The author admits that this kind of strategy is not without risks. People are different and, as Schwartz (2015) remarks, "one patient asked to authorize concealment may feel valued and respected, but another who is asked may simply not know what to think, or may start to question other times the doctor seemed less than completely forthcoming." As a solution to this problem Alfano suggests that the physician could provide the patient with a "bad fortune cookie", a list of the potential side effects in a sealed envelope. The patient could open the envelope if he experienced an unexpected adverse health outcome during the course of treatment, but not otherwise.

Another concept introduced by Alfano is authorized deception, which goes further that authorized concealment. He describes it again with an imaginary example of a doctor talking to a patient:

> Research suggests that you will have better symptomatic outcomes if I lead you to believe some things that are, in fact, false. Naturally, I understand that you want to know everything that's relevant to your prospects, but some of that knowledge might lead you to suffer unnecessarily. What I'd like to do is ask your permission to deceive you about some things in order to promote your welfare. I'll debrief you about them after the treatment is over. (Alfano 2015a)

Also in this case the physician could provide the patient with a "bad fortune cookie", a sealed envelope which the patient could open in case of unexpected problems. In addition, the patient could elect an "agony aunt", "someone who knew the patient well, had somewhat matched values, and could be expected to keep his information confidential." The person would be told the truth about the patient's case, "asked to keep an eye on him, but instructed not to reveal the deception unless it seemed necessary." (Alfano 2015a)

These suggestions are creative, to say the least, but hardly clinically applicable. The burden on the patient to decide about the nature of her symptoms would in most cases be unbearable. How could the patient differentiate between the symptoms that would legitimize opening the envelope and those that would not. Alfano suggests that authorized concealment should be limited to situations where the risks are 'merely symptomatic', but this does not help. Pain, for example, is always 'symptomatic' and it may or may not be a sign of something serious. It is probable that the proposals would create "more stress and nocebo effects than the situation in which the information is presented in the conventional way" (Roeser 2015). A bad fortune cookie and agony aunt seem to be "overly rationalistic solutions to a situation that is fraught much more with emotions and other psychological factors than even Alfano seems to acknowledge" (ibid.).

Meynen and Widdershoven (2015) remark that Alfano does not sufficiently take into account the concrete health care context. According to them, "the efforts by the doctor may be reduced or even nullified when the patient receives information about possible side effects and figures about unsuccessful treatment via the Web." Also, "the patient may lose trust in the physician because of receiving incomplete or otherwise not fully adequate information." (ibid.).

Blease (2015) points out that both 'bad fortune cookies' and an 'agony aunt' might induce or aggravate nocebo effects in patients. She refers to research in cognitive science, according to which secrets are likely to be transmitted to others, in particular when secret disclosures are of intense emotional relevance to the listener (ibid.). According to Blease, "Informed consent in both strategies appears to make a theatrical event out of hidden information, thereby elevating its premium and the desire to unmask it." (ibid.). She makes her own proposal for the doctor to present the possibility of side effects:

> If you take this drug you have a 70 percent chance of feeling better. We can talk about the low risk of side effects if you like but research shows that if we don't dwell on these things, you will have an 80 percent chance of avoiding any unnecessary and unpleasant symptoms. Therefore I'd like to recommend this treatment because of its success rate, and if at any time you need to see me again just make appointment.

Blease's suggestion is a step towards maintaining patient autonomy and simultaneously framing the message in a positive way. It is not practical, however, in the sense that a clinician hardly ever has such numerical data.

Alfanos's ideas about authorized concealment and authorized deception were published as a target article in the American Journal of Bioethics (Alfano 2015a). In a response to the Open Peer Commentaries in the same journal he presented a new idea that might "help to allay autonomy-related concerns about framing, authorized concealment and deception, and the bad fortune cookie and agony aunt solutions" (Alfano 2015b).

His suggestion was the development of 'deep value-matching health apps' that would help patients to find doctors who share their values, not only in health issues but also in moral questions in general (ibid.). The original idea about such matching had been presented by Robert Veatch more than two decades earlier, well before the powerful information technology we have today.

It would be easy to criticize the suggestion for being impractical and utopic. It ignores, for example, the fact that only a small minority of people even in the rich part of the world are in a situation in which they could freely choose their doctors. The role of philosophers is, however, to question current practices and present innovative solutions. In this sense Alfano's idea about 'deep value-matching health apps' is another thought experiment and—considering the pace of technological development during recent years—not even utopic.

4.5 Open Label Placebos

It has been widely assumed that deception of the patient is a necessary condition for placebo effects. A few studies show, however, that this is not the case (Colagiuri et al. 2015; Kaptchuk and Miller 2018).

The first of these studies was carried out already in 1965 and was titled simply "Nonblind placebo trial" (Park and Covi 1965). Fifteen newly admitted neurotic outpatients of a psychiatric clinic were given placebos for one week after being informed as follows:

> Mr. Doe, at the intake conference we discussed your problems and your condition, and it was decided to consider further the possibility and the need of treatment for you before we make a final recommendation next week. Meanwhile, we have a week between now and your next appointment, and we would like to do something to give you some relief from your symptoms. Many different kinds of tranquilizers and similar pills have been used for conditions such as yours, and many of them have helped. Many people with your kind of condition have also been helped by what are sometimes called 'sugar pills', and we feel that a so-called sugar pill may help you, too. Do you know what a sugar pill is? A sugar pill is a pill with no medicine in it at all. I think this pill will help you as it has helped so many others. Are you willing to try this pill? (ibid.)

Fourteen patients completed the study and all but one reported significant improvement in their symptoms. There was also marked improvement by doctor ratings on several measures. At the end of the study, at least five patients decided to continue the placebo treatment and two felt they did not need treatment any more.

According to the authors, the primary finding was that "patients can be willing to take placebo and can improve despite disclosure of the inert content of the pills; belief in pill as drug was not a requirement for improvement."

The conclusions that can be drawn from this pioneering study are, however, limited. The sample size was small and there was no control group in the study. In particular, there was no no-intervention group and we do not know what would have happened to the patients without any intervention.

For over 40 years the study by Park and Covi remained the only one in which placebos were given openly to the patients. The next landmark open-label studies with placebos were conducted in 2008 by Sandler and Bodfish and in 2008 by Kaptchuk et al.

Sandler and Bodfish (2008) performed a prospective crossover trial in 26 children with ADHD. The patients were followed at a community-based ADHD clinic and were stable on stimulant therapy.

They were randomly assigned to one of two orders of experimental conditions:

> (1) a baseline condition during which they received 100% of their current dose of stimulant medication (1 week), then a 50% dose condition during which they received half of their baseline dose (1 week), and finally a 50% dose plus distinctive placebo condition in which they received half their baseline dose and a visually distinctive placebo capsule (1 week);
> (2) a baseline condition during which they received 100% of their current dose of stimulant medication (1 week), then a 50% dose plus distinctive placebo condition in which they received half their baseline dose and a visually distinctive placebo capsule (1 week), and finally a 50% dose condition during which they received half of their baseline dose (1 week).

The inert nature of the placebo was fully disclosed to parent and child. The parents were informed as follows:

> You are being asked to allow your child to participate in a study to examine if children with ADHD can be maintained on a lower dose of stimulant medication with the same level of symptom control and a lowered risk of medication side effects. This may be possible by taking an additional capsule. This capsule is a placebo (a pill containing no active drug or medication), which we think may act as a 'Dose Extender'. (ibid.)

According to the authors, the most important finding of the study was that "ADHD behaviour tended to remain the same when the dose of stimulant medication was reduced with placebo but to deteriorate when the dose was reduced without placebo." (ibid.).

The results are certainly encouraging but at the end of their paper Sandler and Bodfish demonstrate misunderstanding of the complex relationship between the concepts placebo and placebo effect. They write: "we believe this line of research involving therapeutic uses of placebo effects, administered ethically to treat children with lower cumulative doses of stimulants, may hold great potential to improve health care" (ibid.). In their study, the use of placebos was associated with a placebo effect but, as we have seen, use of placebos is not a necessary condition for establishing placebo effects. And to talk about "use of placebo effect" is misleading in the sense that placebo effects are always present in a doctor-patient relationship.

Another innovative and influential study was that of Kaptchuk et al. (2008), who compared standard care with and without the addition of open-label placebo treatment for irritable bowel syndrome (IBS). All patients were informed about the nature of placebos as follows:

placebo pills, something like sugar pills, have been shown in rigorous clinical testing to
produce significant mind-body self-healing processes. (ibid.)

As the text shows, the placebo effect was deliberately framed positively. Patients were
then told that half would be assigned to an open-label placebo group and the other
half to a no-treatment control group. At the end of the interview, patients randomized
to the open-label placebo group were given a typical medicine bottle of placebo pills
with a label'placebo pills—take 2 pills twice daily'. Patients in the no-treatment
group were reminded of the importance of their group for the study.

Altogether 70 patients completed the study. The main result was that IBS symptoms improved in both groups but significantly more in the open-label placebo
group.

The authors discuss carefully the limitations and potential biases of the study
and describe it as a 'proof-of-principle' pilot study because of the small sample
size and short duration. A specific source of bias in this study setting was potential
disappointment due to assignment to the no-treatment group. However, according to
interviews with the patients, 76% of those in the no-treatment group reported they
were not disappointed with their assignment.

Later open-label placebo trials have followed and modified the protocol of the IBS
study by Kaptchuk et al. (2008). Positive results have been obtained, for example,
in chronic low back pain (Carvalho et al. 2016), cancer related fatigue (Hoenemeyer
et al. 2018), episodic migraine attacks (Kam-Hansen et al. 2014) and allergic rhinitis
(Schaefer et al. 2018).

The ethics of disclosure in the open-label placebo trials

Institutional review of research did not exist during the time of the study by Park and
Covi in 1965, but the study protocols by Sandler and Bodfish and Kaptchuk et al.
were approved by the respective review boards of the corresponding institutions. The
patients in all of these studies were openly told that they might end up in a placebo
group and they were also informed about the nature of placebos.

The patients were not deceived but this is not the end of the story. Let us look at
the patient information again:

Many people with your kind of condition have also been helped by what are sometimes
called 'sugar pills' (Park and Covi 1965)

This capsule is a placebo (a pill containing no active drug or medication), which we think
may act as a 'Dose Extender'. (Sandler and Bodfish 2008)

placebo pills, something like sugar pills, have been shown in rigorous clinical testing to
produce significant mind-body self-healing processes. (Kaptchuk et al. 2008)

The language used is common in the placebo literature, but the expressions are only
partly true. In the first case people have actually *not* been helped by placebos but
by the context and the relationship *including* placebos. In the second case, 'Dose
Extender' is a name invented by the researchers to boost the placebo effect. And
in the third case, placebo pills *as such* have not been shown to produce anything.
'Mind-body' and 'self-healing' sound like buzzwords used also to boost the placebo
effect.

In addition, it was misleading to refer to subjects who responded to placebo pills in past trials since those patients did not know the pills were placebos. According to Justman (2013), "an element of concealment was grandfathered into this study of placebos without concealment." He concludes that, "An appropriate response to this dilemma is first of all to consider the placebo effect as a therapeutic benefit arising from the conscientious performance of the rituals of good medicine, and not as a resource to be tapped by the use of trickery (with equivocations counting as trickery) or dispensed in the form of pills." (Justman 2013)

Annoni and Miller (2016) admit that Justman's critique has the merit of "underscoring how subtle the boundary between truthfulness, manipulation, and deception can be" but still claim that it is possible to administer open-label placebos and provide truthful information, for example in the following way:

> This pill is a placebo; as such, it does not contain any pharmacologically active ingredient. It is something like a sugar pill. However, rigorous clinical testing has shown that even this kind of intervention may have clinically relevant effects by inducing what is called a 'placebo response.' Placebo responses are part of the way in which our body and mind reacts to the provision of care and to salient features of the therapeutic context. Thus, if you regularly take these pills, you may experience some improvement in your condition.

The authors have modified the message but it is, in my opinion, still not completely truthful. At least in the ears of the patient "this kind of intervention" may refer to the placebo pills only, in which case the sentence is not true.

The questionable nature of claims like "placebos have been shown to work in this condition" is obvious when 'placebo' is changed for, e.g., 'penicillin'. If "penicillin has been shown to work in this condition", it means that penicillin *as such* has increased the probability of a favourable outcome. A similar claim is never true of placebos even if the claim is framed with valid assertions about placebo effects.

References

Alfano, M. 2015a. Placebo effects and informed consent. *American Journal of Bioethics* 15: 3–12.

Alfano, M. 2015b. Response to open peer commentaries on "Placebo Effects and Informed Consent". *American Journal of Bioethics* 15: W1–W3.

Altman, D.G., and J.M. Bland. 1995. Absence of evidence is not evidence of absence. *BMJ* 311: 485.

AMA (American Medical Association). Use in clinical practice. https://www.ama-assn.org/delivering-care/use-placebo-clinical-practice. Accessed 1 Jun 2018.

Annoni, M., and F.G. Miller. 2016. Placebo effects and the ethics of therapeutic communication: A pragmatic perspective. *Kennedy Institute of Ethics Journal* 26: 79–103.

Blease, C. 2015. Authorized concealment and authorized deception: Well-intended secrets are likely to induce nocebo effects. *American Journal of Bioethics* 15: 23–25.

Bok, S. 2002. Ethical issues in use of placebo in medical practice and clinical trials. In *The science of the placebo: Toward an interdisciplinary research agenda*, ed. H.A. Guess, A. Kleinman, J.W. Kusek, and L.W. Engel, 53–74. London: BMJ Books.

Brown, W.A. 1998. The placebo effect. *Scientific American* 278 (1): 90–95.

Carlino, A. 2005. Petrarch and the early modern critics of medicine. *Journal of Medieval and Early Modern Studies* 35: 559–582.

Carvalho, C., J.M. Caetano, L. Cunha, et al. 2016. Open-label placebo treatment in chronic low back pain: A randomized controlled trial. *Pain* 157: 2766–2772.

Cheng, Maria. 2013. Many UK doctors give useless drugs, treatments. Times Colonist/Associated Press, March 20. http://www.timescolonist.com/life/health/many-uk-doctors-give-patients-useless-drugs-treatments-authorities-say-that-s-unethical-1.94966. Accessed 1 Jun 2018.

Colagiuri, B., L.A. Schenk, M.D. Kessler, et al. 2015. The placebo effect: From concepts to genes. *Neuroscience* 307: 171–190.

Ende, J., L. Kazis, A. Ash, and M. Moskowitz. 1989. Measuring patients' desire for autonomy: Decision-making and information-seeking preferences among medical patients. *Journal of General Internal Medicine* 4: 23–30.

Evans, H.M., and A.P.S. Hungin. 2007. Uncomfortable implications: Placebo equivalence in drug management of a functional illness. *Journal of Medical Ethics* 33: 635–638.

Feder, K.E. 2007. Beyond autonomy and paternalism: The caring transparent self. In *Autonomy and paternalism. between independence and good intentions*, ed. T. Nys, Y. Denier, and T. Vandevelde, 1–29. Leuven: Peeters.

Fent, T., T. Rosemann, M. Fässler, O. Senn, and C.A. Huber. 2011. The use of pure and impure placebo interventions in primary care—A qualitative approach. *BMC Family Practice* 12: 11.

Fässler, M., M. Gnadinger, T. Rosemann, and N. Biller-Andorno. 2009. Use of placebo interventions among Swiss primary care providers. *BMC Health Serv Res* 9: 144.

Fässler, M., K. Meissner, A. Schneider, and K. Linde. 2010. Frequency and circumstances of placebo use in clinical practice—A systematic review of empirical studies. *BMC Medicine* 8: 15.

Gold, H., D.B. Barr, M. Cattell, E.F. DuBois, P.A. Bunn, and W. Modell (eds.). 1947. *Cornell conferences on therapy: Use of placebos in therapy*. New York: Macmillan.

Hemilä, H. 2013. Vitamin C may alleviate exercise-induced bronchoconstriction: A meta-analysis. *BMJ Open* 3: e002416.

Hemilä, H., and Elizabeth Chalker. 2013. Vitamin C for preventing and treating the common cold. *Cochrane Database Systematic Reviews* (1): CD000980.

Hoenemeyer, T.W., T.J. Kaptchuk, T.S. Mehta, et al. 2018. Open-label placebo treatment for cancer-related fatigue: A randomized-controlled clinical trial. *Scientific Reports* 8: 2784.

Holley, A.D., E. Osland, J. Barnes, A. Krishnan, and J.F. Fraser. 2011. Scurvy: Historically a plague of the sailor that remains a consideration in the modern intensive care unit. *Internal Medicine Journal* 41: 283–285.

Howick, J., F.L. Bishop, C. Heneghan, J. Wolstenholme, S. Stevens, F.D.R. Hobbs, et al. 2013. Placebo use in the United Kingdom: Results from a national survey of primary care practitioners. *PLoS ONE* 8 (3): e58247.

Hróbjartsson, A., and M. Norup. 2003. The use of placebo interventions in medical practice: A national questionnaire survey of Danish clinicians. *Evaluation and the Health Professions* 26: 153–165.

Hróbjartsson, A. 2008. Clinical placebo interventions are unethical, unnecessary and unprofessional. *Journal of Clinical Ethics* 19: 66–69.

Juraschek, S.P., E. Guallar, L.J. Appel, and E.R. Miller. 2012. Effects of vitamin C supplementation on blood pressure: A meta-analysis of randomized controlled trials. *American Journal of Clinical Nutrition* 95: 1079–1088.

Justman, S. 2013. Deceit and transparency in placebo research. *Yale Journal of Biology and Medicine* 86: 323–331.

Kaiser, L., D. Lew, B. Hirschel, R. Auckenthaler, A. Morabia, A. Heald, et al. 1996. Effects of antibiotic treatment in the subset of common-cold patients who have bacteria in nasopharyngeal secretions. *Lancet* 347: 1507–1510.

Kam-Hansen, S., M. Jakubowski, J.M. Kelley, et al. 2014. Altered placebo and drug labeling changes the outcome of episodic migraine attacks. *Science Translational Medicine* 6: 218ra5.

Kaptchuk, T.J., J.M. Kelley, L.A. Conboy, R.B. Davis, C.E. Kerr, E.E. Jacobson, et al. 2008. Components of placebo effect: Randomised controlled trial in patients with irritable bowel syndrome. *BMJ* 336: 999–1003.

Kaptchuk, T., and F. Miller. 2018. Open label placebo: Can honestly prescribed placebos evoke meaningful therapeutic benefits? *BMJ* 363: k3889.

Leslie, A. 1954. Ethics and practice of placebo therapy. *American Journal of Medicine* 16: 854–862.

Lichtenberg, P., U. Heresco-Levy, and U. Nitzan. 2004. The ethics of the placebo in clinical practice. *Journal of Medical Ethics* 30: 551–554.

Loewy, E. 2005. In defense of paternalism. *Theoretical Medicine and Bioethics* 26: 445–468.

Louhiala, P. 2009. The ethics of the placebo in clinical practice revisited. *Journal of Medical Ethics* 35: 407–409.

Louhiala, P. 2012. What do we really know about the deliberate use of placebos in clinical practice? *Journal of Medical Ethics* 38: 403–405.

Louhiala, P. 2014. Deciding on treatment. In *The medical humanities companion volume three: Treatment*, ed. P. Louhiala, I. Heath, and J. Saunders, 47–50. Oxford: Radcliffe.

Louhiala, P., H. Hemilä, and R. Puustinen. 2015. Impure placebo is a useless concept. *Theoretical Medicine and Bioethics* 36: 279–289.

Meissner, K., L. Höfner, M. Fässler, and K. Linde. 2012. Widespread use of pure and impure placebo interventions by GPs in Germany. *Family Practice* 29: 79–85.

Meynen, G., and G. Widdershoven. 2015. Dealing with placebo effects: A plea to take into account contextual factors. *American Journal of Bioethics* 15: 19–21.

Nitzan, U., and P. Lichtenberg. 2004. Questionnaire survey on use of placebo. *BMJ* 329: 944–946.

Novella, S. 2013. Do 97% of UK doctors prescribe placebos? Neurologica Blog. http://theness.com/neurologicablog/index.php/do-97-of-uk-doctors-prescribe-placebos/. Accessed 1 Jun 2018.

Nys, T., Y. Denier, and T. Vandevelde. 2007. Introduction. In *Autonomy & paternalism. Reflections on the theory and practice of health care*, ed. T. Nys, Y. Denier, and T. Vandevelde, 1–22. Leeuven: Peeters.

O'Neill, O. 2002. *Autonomy and trust in bioethics*. Cambridge: Cambridge University Press.

Padayatty, S.J., H.D. Riordan, S.M. Hewitt, L.A. Katz, J. Hoffer, and M. Levine. 2006. Intravenously administered vitamin C as cancer therapy: Three cases. *Canadian Medical Association Journal* 2006 (174): 937–942.

Park, L.C., and U. Covi. 1965. Nonblind placebo trial: An exploration of neurotic patients' responses to placebo when its inert content is disclosed. *Archives of General Psychiatry* 12: 36–45.

Pellegrino, E. 2009. Physician integrity: Why it is inviolable. In: Connecting American Values with Health Reform. Hastings Center, 18–20. http://www.thehastingscenter.org/Publications/Detail.aspx?id=3528. Accessed 1 Jun 2018.

Philby, C. 2013. The placebo effect: Doctors admit prescribing unproven treatments, unnecessary tests and pills with no active ingredient. *The Independent*, March 20, 2013. http://www.independent.co.uk/life-style/health-and-families/health-news/the-placebo-effect-doctors-admit-prescribing-unproventreatments-unnecessary-tests-and-pills-with-no-active-ingredient-8542666.html. Accessed 1 Jun 2018.

Raynaud-Simon, A., J. Cohen-Bittan, A. Gouronnec, E. Pautas, P. Senet, M. Verny, and J. Boddaert. 2010. Scurvy in hospitalized elderly patients. *Journal of Nutrition, Health, and Aging* 14: 407–410.

Roeser, S. 2015. Placebo, nocebo, informed consent, and moral technologies. *American Journal of Bioethics* 15: 15–17.

Sandler, A.D., and J.W. Bodfish. 2008. Open-label use of placebos in the treatment of ADHD: A pilot study. *Child Care and Health Development* 34: 104–110.

Schaefer, M., T. Sahin, B. Berstecher, and J.P. van Wouwe. 2018. Why do open-label placebos work? A randomized controlled trial of an open-label placebo induction with and without extended information about the placebo effect in allergic rhinitis. *PLOS ONE* 13(3): e0192758.

ScienceDaily. U.S. doctors regularly prescribe real drugs as placebo treatments, study claims. *ScienceDaily*, October 25, 2008. http://www.sciencedaily.com/releases/2008/10/081023195216. htm. Accessed 1 Jun 2018.

Schneider, C.E. 2005. Some realism about informed consent. *Journal of Laboratory and Clinical Medicine* 145: 289–291.

Schwartz, P.H. 2015. Placebos, full disclosure, and trust: The risks and benefits of disclosing risks and benefits. *American Journal of Bioethics* 15: 13–14.

Shapiro, A., L.A. Morris. 1978. The placebo effect in medical and psychological therapies. In *Handbook and psychotherapy and behavior change: An empirical analysis,* 2nd edn., 369–410, eds. S.L. Garfield, A.E. Bergin. New York: Wiley.

Sherman, R., and J. Hickner. 2008. Academic physicians use placebos in clinical practice and believe in the mind–body connection. *Journal of General Internal Medicine* 23: 7–10.

Tauber, A.I. 2005. *Patient autonomy and the ethics of responsibility.* Cambridge, Mass; London: MIT Press.

Tilburt, J.C., E.J. Emanuel, T.J. Kaptchuk, and F.A. Curlin. 2008. Prescribing 'placebo treatments': Results of national survey of US internists and rheumatologists. *British Medical Journal* 337: a1938.

Wang, Y., X.J. Liu, L. Robitaille, S. Eintracht, E. MacNamara, and L.J. Hoffer. 2013. Effects of vitamin C and vitamin D administration on mood and distress in acutely hospitalized patients. *American Journal of Clinical Nutrition* 2013 (98): 705–711.

Young, J., A. De Sutter, D. Merenstein, G.A. van Essen, L. Kaiser, H. Varonen, et al. 2008. Antibiotics for adults with clinically diagnosed acute rhinosinusitis: A meta-analysis of individual patient data. *Lancet* 371: 908–914.

Zhang, M., L. Robitaille, S. Eintracht, and L.J. Hoffer. 2011. Vitamin C provision improves mood in acutely hospitalized patients. *Nutrition* 27: 530–533.

Chapter 5
Placebo Research and Clinical Practice

> Very soon the discussion revealed–certainly not for the first time in the history of medicine–that by far the most frequently used drug in general practice was the doctor himself. It was not only the medicine in the bottle, or the pills in the box, that mattered, but the way the doctor gave them to his patient–in fact the whole atmosphere in which the drug was given and taken. (Balint 1955)

> It is plausible to argue that research on placebo and nocebo effects may not only prompt a revolutionary shift in thinking of the physician-patient interaction, with the promise to guide strategies for optimizing clinical practice, but will also open promising avenues for improvement within most areas of modern medicine. (Meissner et al. 2011)

> Practitioner characteristics, like empathy, friendliness, and competence, favor the formation of positive expectancies. Caring and warm patient–practitioner interactions can enhance the therapeutic value of clinical encounters when patients' positive expectancies are actively encouraged and engaged. (Blasini et al. 2018)

Placebos–understood as inert substances–were introduced in the late 1940's as an important tool in clinical research. From the beginning, placebo effect has had two meanings: the changes in the placebo group in research and the changes in an individual patient in clinical care. These two situations are fundamentally different, since the aim of research is to find out whether a new method works and the aim of clinical medicine is to help the individual patient. For reasons presented in Chap. 2, 'placebo effect' is a confusing label to describe the changes in a patient in clinical care.

The confusing terminology and the negative connotations related to 'placebo' may explain why the results of empirical research into placebo and nocebo phenomena have received so little attention outside the research community and little has been done to translate the knowledge into improved clinical care (Faria et al. 2014; Lucassen and Olesen 2016).

Di Blasi et al. (2001) published a systematic review on context effects, focusing on doctor-patient relationships. The studies included were heterogeneous but the authors concluded that

© Springer Nature Switzerland AG 2020
P. Louhiala, *Placebo Effects: The Meaning of Care in Medicine*,
International Library of Ethics, Law, and the New Medicine 81,
https://doi.org/10.1007/978-3-030-27329-3_5

one relatively consistent finding is that physicians who adopt a warm, friendly, and reassuring manner are more effective than those who keep consultations formal and do not offer reassurance. (Di Blasi et al. 2001)

Feinstein called for more research in this area in 2002:

> The elements of good therapeutic style are still regularly and effectively used by many excellent practitioners, who have become an endangered species in today's 'evidence-based', clinically reductionist medical world. Before members of that species become extinct, their knowledge and judgment might be discerned by clinical investigators who want to preserve the value of "therapeutic style" in the humane care of patients. (Feinstein 2002)

His call was heard and during recent years the value of the findings of placebo research has been noticed in several reviews. Annoni and Miller (2016) write that

> These studies demonstrate that the way in which health professionals communicate, disclose, frame, and contextualize information to patients may modulate symptoms across an array of highly prevalent conditions such as pain, depression, anxiety, insomnia, irritable bowel syndrome, migraine, and Parkinson's disease. (Annoni and Miller 2016)

According to Lucassen and Olesen (2016),

> The new research shows that it is due time to create a renaissance for the value of the personal doctor, who is well aware that s/he induces effects that can be biomedically detected and clinically measured. (Lucassen and Olesen 2016)

Arnold et al. (2015) point out that placebo and nocebo effects should be destigmatised, since they are an essential ingredient of routine medical practice. We can aim to enhance placebo effects and suppress nocebo effects but they cannot be separated or eliminated from clinical practice. In fact, to talk about 'using placebo effects' is partly misleading since they cannot 'not be used'.

Another neglected area related to research on placebo effects is the role of spontaneous remission. In trials of depression treatment, for example, it is typical that 30–50% of the patients in placebo groups improve. It is easy to understand that the interest of a manufacturer of a new antidepressant focuses on the difference between the results in the drug group and the placebo group, but researchers should also focus on the factors behind spontaneous remission (Cuijpers and Cristea 2015). The effects of both antidepressants and psychological treatments for depression are modest and it would serve the interests of patients to develop methods to optimise spontaneous remission.

5.1   Context Effect

The role of context has been mentioned in this book many times and, as we saw in Chap. 2, 'contextual healing' has been suggested as a reconceptualization of 'placebo effect' (Miller and Kaptchuk 2008). The use of the concept 'contextual healing' has not, however, gained wider usage but the role of contextual factors has been recognised. In a recent review by Rossettini et al. (2018), for example, the term 'contextual factors' (CFs) is deliberately used instead of placebo. According to the authors,

> CFs are a complex set of internal, external or relational elements encompassing: patient's expectation, history, baseline characteristics; clinician's behavior, belief, verbal suggestions and therapeutic touch; positive therapeutic encounter, patient-centered approach and social learning; overt therapy, posology of intervention, modality of treatment administration; marketing features of treatment and health care setting. (ibid.)

Rossettini et al. classify contextual factors into three classes: (1) internal (e.g., memories, emotions, expectations and psychological characteristics of the patient; (2) external (e.g., physical aspects of therapy, such as the kind of treatment (pharmacological or manual) and the place in which the treatment is delivered) and (3) relational (e.g., the social cues that characterize the patient-physiotherapist relationship, such as the verbal information that the therapist gives to the patient, the communication style or the body language) (ibid.).

The largest body of research on context effects/placebo effects exists in the area of pain treatment and the effects have been quantified in a number of conditions, such as osteoarthritis, low back pain, rheumatoid arthritis and fibromyalgia. In one meta-analysis, a moderate effect size (0.53) was found, when placebo groups were compared to no-treatment groups (Chen et al. 2017). Particularly interesting is that the context effect often equals or even exceeds the effect of the specific treatment (Lucassen and Olesen 2016).

5.2 What Can the Physician Do?

The main factor around the patient that leads to context/ placebo effects is the doctor, but the role of other health professionals may be equally important. Physician-patient and other relationships in care matter and this is not only common sense or old-fashioned paternalism but scientific knowledge, based on solid evidence. The open-hidden paradigm provides perhaps the simplest and most elegant evidence for the power of therapeutic interaction: just telling a patient that she will receive a painkiller enhances the effect significantly.

Barrett et al. (2006) have suggested that there are at least two larger processes involved in health-related placebo/ context effects. The first is 'feeling cared for' (or 'being helped' or 'receiving treatment'). The second is 'empowerment' (or 'taking care of one's self' or 'achieving health' or 'self-actualization'). Health care professionals can facilitate these processes in many ways, and Barrett et al. (2006) list eight actions that are potentially useful:

> speak positively about treatments, provide encouragement, develop trust, provide reassurance, support relationships, respect uniqueness, explore values, and create ceremony.

In the following, I shall describe empirical studies addressing the role of the physician. The classification (ensuring a supportive atmosphere/being positive/giving information/shared decision-making) is from the review by Lucassen and Olesen (2016).

Ensuring a supportive atmosphere

Therapeutic alliance, the bond between the health care professional and the patient, has been widely studied in the context of psychotherapy, but it is equally relevant in all areas of clinical medicine. Krupnick et al. (1996) studied the relationship between therapeutic alliance and treatment outcome for depressed outpatients who received either interpersonal psychotherapy, cognitive-behaviour therapy, antidepressant imipramine with clinical management, or placebo with clinical management (clinical management means here visits to the doctor without psychotherapy). Part of the therapy sessions or other visits were videotaped and the therapeutic alliance rated by trained raters on a validated score. It was found that therapeutic alliance had a significant effect on the outcome in all groups, including the imipramine and placebo groups.

Physicians' communication styles have also been studied experimentally in a few trials. Verheul et al. (2010), for example, assessed the effects of physicians' communication style and patients' provoked expectations in thirty healthy subjects in a role-play consultation with a general practitioner (GP). The study was a randomised trial in which the GP communicated in a warm and empathic or cold and formal way and raised positive or uncertain expectations. Before meeting the GP, the 'patients' received a script on the nature of their complaint which was severe menstrual pain. Because this was a role-play, there was no actual pain to be measured. Instead, the study subjects' anxiety, expectation and affects were measured. The main result was that "only warm, empathic communication combined with positive expectations led to a significant and relevant decrease in state anxiety" (ibid.). The authors conclude that generating positive expectations in a patient is not enough, since if this is done in a cold and formal way, patients' anxiety will not decrease. Warm and empathic communication is also insufficient, if positive expectations are not raised simultaneously.

Kaptchuk et al. (2008) conducted a prospective study on the elements of placebo effects (their term in the article although Kaptchuk has elsewhere written about contextual healing). The patients had irritable bowel syndrome (IBS) and were randomly assigned to three groups: (1) waiting list, (2) placebo acupuncture alone ('limited interaction') and (3) placebo acupuncture with a positive patient-practitioner relationship ('augmented interaction'). The placebo treatment in groups 2 and 3 was a validated sham acupuncture device. The outcomes of the study were change from baseline in the global IBS improvement scale, symptom severity scale and quality of life scale. It was found that 28% of the patients on the waiting list, 44% in the limited interaction group, and 62% in the augmented interaction group reported adequate relief of their symptoms. The same trend in response was found in all scales measured.

The authors discussed the limitations of the study at considerable length. They noted that a genuine no-treatment group would have been a waiting-list group in which the participants would have been followed without their knowledge. For operational and ethical reasons, it is, however, difficult to arrange such a control group. Another limitation mentioned was the subjective nature of the outcome measures. This limitation is, of course, present in most research on placebo effects since pain and other subjective symptoms are the usual variables measured in placebo research.

A study on the effect of the consultation process as such was conducted by Brien et al. (2011), who aimed at finding out whether possible benefits from homeopathic intervention in patients with rheumatoid arthritis are due to the homeopathic consultation, homeopathic remedies or both. They performed a trial in which the patients were randomised into five groups. Three groups received a homeopathic consultation and two did not. Patients in the consultation groups were further randomised to receive individualised homeopathic treatment, standardised homeopathic treatment or placebo. Patients in the non-consultation group were randomised to receive standardised homeopathic treatment or placebo. All patients remained on their conventional medication also. It was found that the benefit of homeopathy was attributable to the consultation, not the homeopathic preparations.

Interestingly, the authors write that

> The placebo effects of the homeopathic consultation may well be specific to this therapy possibly being dependent on the ritual of the collaborative and highly individualized consultation necessary to identify a homeopathic remedy and the associated symbolic meaning response for that patient. (ibid.)

This statement is speculative and the study itself does not provide any support for it. I agree with Ernst (2011) who in an accompanying editorial suggests that we simply take the results at face value:

> Homeopathic remedies are ineffective and empathetic therapeutic encounters are helpful. So, we should discard the ineffective and adopt the helpful. (ibid.)

The relationship between empathy in the medical consultation and the outcome of the common cold was assessed in a study by Rakel et al. (2009). Patients with new symptoms were randomised to "(1) no practitioner-patient interaction, (2) 'standard' practitioner-patient interaction, and (3) 'enhanced' practitioner-patient interaction". The latter included five added ingredients of the practitioner-patient interaction (positive prognosis, empathy, empowerment, connection, and education). It was found that when patients rated encounters with clinicians as "perfect" on a measure of empathy, they had a shorter duration of illness (7.1 vs. 8.0 days) and a trend toward lesser severity of illness (16% lower symptom score).

Being positive

A classical study on the role of being positive was conducted by Thomas in 1987. The study population consisted of 200 general practice patients with symptoms but no abnormal physical signs and for whom no definite diagnosis could be made. The symptoms the patients had were, for example, cough, sore throat, abdominal pain, back pain, headache and tiredness. The patients were then randomised to one of four consultations: "a consultation conducted in a 'positive manner', with and without treatment, and a consultation conducted in a 'non-positive manner', called a negative consultation, with and without treatment" (Thomas 1987). In a 'positive consultation' "the patient was given a firm diagnosis and told confidently that he would be better in a few days". In a negative consultation "no firm assurance was given" and the GP could say, for example: "I cannot be certain what is the matter with you". The 'treatment' was a prescription for a small dose of thiamine (vitamin B1), which in this case was not expected to have a significant biological effect on the symptoms.

The study was conducted over 30 years ago and, judging by today's standards, was methodologically and ethically problematic in many ways: it is questionable to lump together patients with such a large variety of symptoms; obviously the patients' informed consent was not obtained, the same GP was both providing the consultations and evaluating the results etc. The same critique can, however, be presented against any old study, and it does not undermine the relevance of the results. The main finding of the study was that after two weeks, 64% of those who received a positive consultation got better compared with 39% of those who received a negative consultation. Treatment with thiamine did not make a difference. The author concluded that "possibly the doctor himself is still the most effective treatment available".

As we have seen earlier, patients' expectations of the effectiveness of treatments explain a major part of placebo and nocebo effects. The evidence has accumulated from experimental research with inert compounds, and little research has been done with real drugs. Bingel et al. (2011) investigated the role of expectancies in the effect of remifentanil, which is a potent opioid. The subjects were healthy volunteers who were exposed to experimental heat pain. It was found that positive treatment expectancy doubled the analgesic benefit of remifentanil and negative treatment expectancy abolished the effect of the drug (ibid.).

Another example of the role of positive or negative framing of information is a study on the reported side effects and work absenteeism related to influenza vaccination (O'Connor et al. 1996). Nearly 300 unimmunized patients with a history of chronic respiratory or cardiac disease were randomised to receive either positively or negatively framed information about the side effect of the vaccine. Framing did not influence the patients' decisions to be vaccinated but the positive frame group reported fewer side effects and less work absenteeism. It cannot be known if the negative frame group actually had more side effects or were simply more aware of these effects due to the framing. Nevertheless, the significant difference in work absenteeism is a clinically relevant finding.

Recent research on mindsets (see Sect. 3.3.) has also confirmed the importance of positive physician-patient relationships. Mindsets,

> like beliefs, guide attention and motivation in ways that shape physiology and behavior; they are related to but distinct from heuristics, which are mental shortcuts used to make decisions under uncertainty and allow individuals to make decisions quickly and efficiently

to solve problems. Sometimes grounded in facts and sometimes not, mindsets are biased or simplified versions of what is right, natural, or possible. (Crum and Zuckerman 2017)

Mindsets are influenced by, for example, culture, media and other people---as are physicians. In the medical context, mindsets concerning the effects of treatments and mindsets about the capacity to change are particularly important (ibid.). An example of the latter is the patient's broader mindset about the nature of health. If the patient believes that genes determine body weight, she is probably less motivated to start to diet or exercise. Physicians may have an important role in promoting 'growth mindsets', according to which patients have inner resources for change.

Giving information

The first experimental study about the effect of encouragement and education of patients was performed in 1964 by Egbert et al. (1964) for a group of 97 patients who had elective intra-abdominal operations (e.g., cholecystectomy, bowel resection or hysterectomy). Before surgery, the patients were randomly assigned to two groups. The 'special care' patients were told what to expect postoperatively and they were taught how to relax, how to take deep breaths and how to move to remain comfortable after the operations. All information was given "in a manner of enthusiasm and confidence" (ibid.). Patients in the control group were not told about the postoperative pain. The surgeons did not know which patients were receiving 'special care' and the patients did not know that they were included in a study.

After the day of the operation, patients in the 'special care' group needed about 50% less pain medication than patients in the control group. They were also discharged from the hospital on average 2.7 days earlier than the control group. The authors concluded that

if an anesthetist considers himself a doctor who alleviates pain associated with operations, he must realize that only part of his work is in the operating room; the patients need ward care by their anesthetists as well. (ibid.)

This conclusion is perhaps self-evident from today's perspective. The study was, however, an important forerunner although it is not possible to infer the weight of the different parts of the intervention in the 'special care' group.

The existing research on the effect of physician-patient communication has been summarised in two review articles, one over 20 years ago (Stewart 1995) and the other more recently (Weiland 2012). Stewart (1995) found 21 randomised trials or analytic studies of physician-patient communication in which the outcome was the patient's health status. The outcomes measured were, for example, symptom resolution, functioning, pain control, emotional health and some physiologic measures like blood pressure. She found that most of the studies "demonstrated a correlation between effective physician-patient communication and improved patient health outcomes" (ibid.).

The review by Weiland (2012) concentrated on studies addressing the role of communication in patients with medically unexplained physical symptoms (MUPS). They found that, in general, patients report fewer symptoms and health anxiety when symptoms are properly explained. A positive physician-patient relationship and positive feedback from the doctor lead to better coping with long-term symptoms and fewer contacts with health care (ibid.).

Shared decision-making

In general, patients have taken (or they have been given) a more active role in treatment decisions during the past five decades. This is mainly a result of the general development of patient autonomy into a key value in health care. Concepts like *patient-centredness* or *shared decision-making* illustrate the development.

Empirical research on placebo and nocebo effects has, for its part, demonstrated that involving patients in decisions may enhance their expectations and in this way also improve the outcome.

Bartley et al. (2016), for example, studied the impact of having a choice of medication compared to no choice in 61 healthy students. The condition examined was pre-examination anxiety. The participants were randomized to two groups: in group 1 they could choose between two study drugs and in group 2 they were assigned to one of them. They were told the drugs were beta-blockers that were equally effective and had similar side-effect profiles. The study drugs were, in fact, placebos.

There were significant differences between the groups with respect to both placebo and nocebo effects. Participants in the choice group showed a reduction and those in the no-choice group an increase in heart rate when they were given a stressful task. The no-choice group reported more side effects than the choice group. The authors remark that the choice provided to the participants was actually superficial and probably the effect of choice on placebo and nocebo effects would have been even more striking if the choice had been between different treatments or if the individuals had had preferences for one treatment over the other.

Geers et al. (2013) conducted four separate studies to test the hypothesis that choice over treatment alternatives leads to a better outcome in experimental conditions. The participants were exposed to laboratory pain or to aversive sounds and their treatment beliefs, treatment choices and their differences in the desire for control were manipulated. The authors concluded that

> The results of the present studies indicate that when individuals exercise choice over treatment alternatives, this choice can strengthen the psychological component of treatment effects for individuals desiring greater control. (ibid.)

It is worth noting here that exercising choice over treatment alternatives enhanced treatment outcomes in those people who had a *high desire of control* (ibid.). The situation is probably different in people who do not wish to control their treatment decisions.

There is no physician without a health care system

No physician is an island and the physician-patient relationship takes place within an organization that is part of a political system of a particular country. Therefore, how well an individual GP or other doctor can harness the positive care effect, depends not only on herself but also on circumstances in which she works. Iona Heath, a London GP for 35 years and the past president of the Royal College of General Practitioners, warned about some tendencies that are not only local in the UK but also global: "Machines have displaced listening and touch, numbers have displaced words, and care and kindness are slowly leaching out of the system" (Heath 2018).

References

Annoni, M., and F.G. Miller. 2016. Placebo effects and the ethics of therapeutic communication: A pragmatic perspective. *Kennedy Institute of Ethics Journal* 26: 79–103.

Arnold, M.H., D.G. Finniss, and I. Kerridge. 2015. Destigmatising the placebo effect. *American Journal of Bioethics* 15 (10): 21–23.

Balint, M. 1955. The doctor, his patient, and the illness. *Lancet* 268: 683–688.

Barrett, B., D. Muller, D. Rakel, D. Rabago, L. Marchand, and J.C. Scheder. 2006. Placebo, meaning and health. *Perspectives of Biology and Medicine* 49: 178–198.

Bartley, H., K. Faasse, R. Horne, and K.J. Petrie. 2016. You can't always get what you want: The influence of choice on nocebo and placebo responding. *Annals of Behavioral Medicine* 50: 445–451.

Bingel, U., V. Wanigasekera, K. Wiech, R. Ni Mhuircheartaigh, M.C. Lee, M. Ploner, et al. 2011. The effect of treatment expectation on drug efficacy: Imaging the analgesic benefit of the opioid remifentanil. *Science Translational Medicine* 3: 70ra14.

Blasini, M., N. Peiris, T. Wright, et al. 2018. The role of patient–practitioner relationships in placebo and nocebo phenomena. *International Review of Neurobiology* 139: 211–231.

Brien, S., L. Lachance, P. Prescott, et al. 2011. Homeopathy has clinical benefits in rheumatoid arthritis patients that are attributable to the consultation process but not the homeopathic remedy: A randomized controlled trial. *Rheumatology* 50: 1070–1082.

Chen, X., K. Zou, N. Abdullah, N. Whiteside, A. Sarmanova, M. Doherty, and W. Zhang. 2017. The placebo effect and its determinants in fibromyalgia: Meta-analysis of randomised controlled trials. *Clinical Rheumatology* 36: 1623–1630.

Crum, A., and B. Zuckerman. 2017. Changing mindsets to enhance treatment effectiveness. *JAMA* 317: 2063–2064.

Cuijpers, P., and I.A. Cristea. 2015. What if a placebo effect explained all the activity of depression treatments? *World Psychiatry* 14: 310–311.

Di Blasi, Z., E. Harkness, E.W. Ernst, A. Georgiou, and J. Kleijnen. 2001. Influence of context effects on health outcomes: A systematic review. *Lancet* 357: 757–762.

Egbert, L.D., G.E. Battit, C.E. Welch, and M.K. Bartlett. 1964. Reduction of postoperative pain by encouragement and instruction of patients. *New England Journal of Medicine* 270(16): 825–827.

Ernst, E. 2011. Homeopathy, non-specific effects and good medicine have we lost core medical values? *Rheumatology* 50: 1007–1008.

Faria, V., C. Linnman, A. Lebel, and D. Borsook. 2014. Harnessing the placebo effect in pediatric migraine clinic. *Journal of Pediatrics* 165: 659–665.

Feinstein, A.R. 2002. Post-therapeutic response and therapeutic 'Style': Re-formulating the 'Placebo Effect'. *Journal of Clinical Epidemiology* 55: 427–429.

Geers, A.L., J.P. Rose, S.L. Fowler, H.M. Rasinski, J.A. Brown, and S.G. Helfer. 2013. Why does choice enhance treatment effectiveness? Using placebo treatments to demonstrate the role of personal control. *Journal of Personality and Social Psychology* 105: 549–566.

Heath, I. 2018. Back to the future: Aspects of the NHS that should never change. *BMJ* 362: k3187.

Kaptchuk, T.J., J.M. Kelley, L.A. Conboy, R.B. Davis, C.E. Kerr, E.E. Jacobson, et al. 2008. Components of placebo effect: randomised controlled trial in patients with irritable bowel syndrome. *BMJ* 336(7651): 999–1003.

Krupnick, J.L., S.M. Sotsky, S. Simmens, J. Moyer, I. Elkin, J. Watkins, et al. 1996. The role of the therapeutic alliance in psychotherapy and pharmacotherapy outcome: Findings in the National

Institute of Mental Health Treatment of Depression Collaboration Research Program. *Journal of Consulting and Clinical Psychology* 64 (3): 532–539.

Lucassen, P., and F. Olesen. 2016. Context as a drug: Some consequences of placebo research for primary care. *Scandinavian Journal of Primary Health Care* 34: 428–433.

Meissner, K., N. Kohls, and L. Colloca. 2011. Introduction to placebo effects in medicine: Mechanisms and clinical implications. *Philosophical Transactions of the Royal Society B* 366: 1783–1789.

Miller, F.G., and T.J. Kaptchuk. 2008. The power of context: Reconceptualizing the placebo effect. *Journal of the Royal Society of Medicine* 101: 222–225.

O'Connor, A.M., R.A. Pennie, and R.E. Dales. 1996. Framing effects on expectations, decisions, and side effects experienced: The case of influenza immunization. *Journal of Clinical Epidemiology* 49: 1271–1276.

Rakel, D.P., T.J. Hoeft, B.P. Barrett, B.A. Chewning, B.M. Craig, and M. Niu. 2009. Practitioner empathy and the duration of the common cold. *Family Medicine* 41: 494–501.

Rossettini, G., E. Carlino, and M. Testa. 2018. Clinical relevance of contextual factors as triggers of placebo and nocebo effects in musculoskeletal pain. *BMC Musculoskeletal Disorders* 19: 27.

Stewart, M.A. 1995. Effective physician-patient communication and health outcomes: A review. *Canadian Medical Association Journal* 152: 1423–1433.

Thomas, K.B. 1987. General practice consultations: Is there any point in being positive? *BMJ* 294: 1200–1202.

Verheul, W., A. Sanders, and J. Bensing. 2010. The effects of physicians' affect-oriented communication style and raising expectations on analogue patients' anxiety, affect and expectancies. *Patient Education and Counseling* 80: 300–306.

Weiland, A. 2012. Encounters between medical specialists and patients with medically unexplained physical symptoms; influences of communication on patient outcomes and use of health care: A literature overview. *Perspectives on Medical Education* 1: 192–206.

Chapter 6
Conclusions

There is a lot of confusion around the concepts 'placebo' and 'placebo effect' both among lay people and the medical profession. The primary aim of this book has been to make sense of these concepts and the phenomena they refer to. In the first chapter I introduced the problems and most of the following chapters have elaborated these issues in the light of empirical research and conceptual analysis. Here, I shall shortly summarise the findings.

The term placebo turned out to be far more complex than the inactive substance of the common definition. Even in the scientific context it is common for placebo to refer to the context of care more broadly and it is sometimes used to cover practically all aspects of the clinician-patient encounter, except for the specific effect of a drug or other treatment.

A major cause of misleading interpretations of 'placebo effect' is the negative language related to 'placebo'. If a placebo is 'inert', 'dummy' or 'sham', these negative connotations are transferred into 'placebo effect', although in the clinical context the latter is a positive and welcomed phenomenon.

From a strictly biological point of view, placebos–like any other substances in nature—are not *fully* inert. The placebos used in clinical research are, in practice and in most cases, inert enough, but sometimes substances like vitamin C or lactose are used as placebos and this may interfere with the study process.

The popular claim about the wide use of placebos in clinical practice turned out to be false. Placebos, understood as (practically) inert substances, are very rarely used in clinical medicine. The misconception about the wide use of placebos is based on the poorly defined concept 'impure placebo' which is of little value and should not be used in scientific or medical literature.

The conceptual confusion is common also in discussions concerning the justification of the clinical use of placebos. Those who support such use seem to refer to 'placebo' in a broader sense than the usual meaning. Sometimes they refer to the 'use of a placebo effect' which is a problematic expression in the sense that 'placebo effect' is not another method that could be used or not used. 'Placebo effect' cannot

© Springer Nature Switzerland AG 2020
P. Louhiala, *Placebo Effects: The Meaning of Care in Medicine*,
International Library of Ethics, Law, and the New Medicine 81,
https://doi.org/10.1007/978-3-030-27329-3_6

be separated from the clinical context. Research on so-called open-label placebos is going on but it is premature to say anything about its clinical implications.

Placebo effects are real and often powerful. There is not, however, a phenomenon that could be labelled 'the powerful placebo' that has a "high degree of therapeutic effectiveness in treating subjective responses" as Henry Beecher wrote in his classical paper in 1955. And contrary to the common claims, Hrobjartsson and Gøtzsche never wrote in their, also classical, paper in 2002, that placebo effects do not exist. Their interest was to explore the potential clinical effects of placebos and they stated explicitly that they did not review the effects of the patient-provider relationship.

There is no necessary connection between the use of placebos and the occurrence of placebo effects. Studies using the *open-hidden paradigm*, for example, demonstrate elegantly that placebos are not needed to provoke placebo effects.

The worlds of clinical research and clinical practice are apart and we should be careful when drawing practical conclusions from research. The aim of a scientific study is to increase our knowledge and the aim of clinical practice is to help patients. Therefore, a certain response in a placebo group of a study does not imply a similar response in the clinical context.

A *general* placebo personality does not exist. It is, however, possible that genes or personality traits will be found that, in *certain circumstances*, increase the probability of a *specific form* of placebo effect. This may increase our understanding of the nature of placebo effects in clinical practice and in the research context.

Nocebo effects are as real and powerful as placebo effects. As in the case of placebo effects, expectations play a significant role in their occurrence. An important clinical example is the occurrence of treatment-related side-effects that are not causally attributable to the treatment as such.

If we use the dichotomy between physiological and psychological, placebo effects belong to both worlds. Major psychological mechanisms as well as neurobiological correlates to them have been described in detail and research is making progress on both fronts. We are used to conceptual frameworks in which mind and body are considered separate but that is also a major source of controversy and confusion around concepts like placebo effect or nocebo effect. It may well be that the mind–body duality is fundamentally false and we simply do not have language to describe the phenomena in a coherent way.

Medical concepts have a past, a present and a future, and often they are replaced with new ones that depict the world more truthfully and do more justice to the phenomena the concepts refer to. In this book, I have described the past and the present of 'placebo' and 'placebo effect' and referred to the common misunderstandings these concepts create. It may take time, but I believe that someday they will be replaced with other terms that create less confusion.

So far, two kinds of suggestions have been made to solve the problems related to 'placebo' and 'placebo effect'. On the one hand, it has been suggested that we abandon the concepts altogether. In the clinical trial context, experiments would simply compare something with something else, a new medication with lactose tablets or an operation with a pseudo-operation, for example. In the clinical practice context there is no place for inert treatments and 'placebo effect' is a misleading name for the effect of the physician-patient relationship and the context.

On the other hand, several attempts have been made to clarify the issues by renaming 'placebo effect'. In my opinion, 'positive care effect', 'contextual healing' and 'meaning response' are the most promising new suggestions although none of them has been widely acknowledged.

Care effect was originally suggested as an alternative to placebo effect in the clinical context. It does not contain the word 'placebo' with its pejorative connotations. Care effect as such could refer to both positive and negative outcomes and therefore the terms *positive* and *negative* care effect would be more logical alternatives to placebo and nocebo effects, correspondingly.

The role of *context* in the clinical encounter is enormous, particularly in chronic conditions like pain. *Contextual healing* is a term that refers to, e.g., the rituals of treatment, the environment, and the communication between the patient and the clinician. Like care effect, contextual healing avoids the current negative terminology.

Meaning response refers to the physiological or psychological effects of meaning in the origins or treatment of illness. Life, in general, is full of meanings and most elements of the practice of medicine are, in fact, meaningful, whether or not the physician or the patient pays attention to them. In the medical context, a positive and negative meaning response would roughly correspond to what has been called placebo effect and nocebo effect.

A simple example of meaning in medicine is the power of diagnoses. The patient enters a physician's office with vague pain but leaves with a diagnosis that may change her whole life although not much has changed in the patient's body. With a diagnosis often comes a prognostic judgment that may influence health outcomes far beyond the bodily processes involved. The meaning of the patient's symptoms may change dramatically also when the physician finds out that the patient does *not* have the disease she's been afraid of.

The limits of our language are obvious when we try to understand how our minds and bodies interact. In fact, our language seems to presume that there is a mind which can act on the body, but that may be a wrong presumption. In the end, subjective responses are not two processes—one mental and one bodily—but a single process, for which, at least so far, science has no words.

One of the tasks of philosophy is to clarify the concepts we use. In our case, *phenomenology* seems to offer perspectives that help us understand the complex phenomena commonly referred to as placebo effects. Phenomenology focuses on the *lived experience* of people: I am not a mind and a body, but an *embodied subjectivity*. Everything is physical and biological but also social, emotional, purposeful and meaningful.

Words matter, relationships matter and circumstances matter. In other words, meanings created in the healing context are of crucial importance for the outcome.

Research on placebo effects has reminded us of the essence of medicine. The key concept is not science but care. Medicine is basically not a science but a humanistic project which uses science to pursue its ends. The effect of positive care effect and the context of care on patients' symptoms and well-being lies on a solid scientific basis and it is unethical to ignore this effect.

The manufacturer's authorised representative in the EU is Springer
Nature Customer Service Centre GmbH, Europaplatz 3, 69115 Heidelberg,
Germany. If you have any concerns regarding our products, please
contact ProductSafety@springernature.com

Printed and bound by CPI Group (UK) Ltd, Croydon, CR0 4YY
29/04/2026
02099460-0014